First paperback edition October 2021

ISBN 978-1-7340696-5-5 (paperback)

Published by Eltahir Research Group

NATURAL VARIABILITY OF THE NILE FLOODS

From Pharaoh's Dream to El Niño / La Niña

BY

ELFATIH A. B. ELTAHIR

Breen M. Kerr Professor of Hydrology and Climate

Professor, Department of Civil and Environmental Engineering

Massachusetts Institute of Technology

with contributions from

GUILING WANG

MOHAMED S. SIAM

MODATHIR A. H. ZAROUG

Table of contents

Edited by Stephanie M. McPherson

Front cover images: Average SST anomalies 9 Aug 2020 – 5 Sept 2020, credit NOAA; British Museum, Room 4: Colossal red granite statue of Amenhotep III, 1350 BC

NATURAL VARIABILITY OF THE NILE FLOODS

Chapter 1—Introduction

Elfatih A. B. Eltahir

Natural variability in the annual flow of the Nile River has been the subject of great interest to the civilizations that have historically occupied the banks of that river. The stories in the Bible and the Koran about prophet Joseph and the dream of the Pharaoh of Egypt point, among other things, to the importance of the Nile water to the ancient civilizations of Egypt. These stories tell of a prediction regarding a period of food surplus followed by a famine episode in Egypt.

In recent history, the High Aswan Dam was constructed to control the annual flow of the Nile and to protect the region against droughts and floods that have clearly affected its people for millennia. The huge structure of the dam stands as a witness to the desperate efforts by man to control the natural variability of this vital resource. Any reduction in the uncertainty about the Nile flood would represent a valuable contribution to water resources management in Egypt. This is particularly true given the natural and social conditions in this arid region, where water is by far the most important natural resource, and where the current population problem will increase the demand on water beyond the generous supply of the Nile.

In Chapter 2, we present results from analysis on two extensive data sets describing sea surface temperature (SST) of the Pacific Ocean, and the flow of water in the Nile River. The analysis suggests that 25% of the natural variability in the annual flow of the Nile is associated with El Niño oscillations. A procedure is developed for using this observed correlation to improve the predictability of the Nile flood. A simple hypothesis is presented to explain physically the occurrence of the Hurst phenomenon in the Nile flow.

Nilometers have been used for gauging the level of water in the Nile river for more than five millennia. The written records describing some of these measurements represent the longest written records for any hydrological phenomenon. They describe interannual fluctuations in the Nile river flow which are closely associated with El Niño phenomenon. In Chapter 3, we use information about long-term variability in El Niño occurrences that has been extracted from the Nilometers records to test the significance of the recent trend in the frequency of El Niño years. We show that the observed frequency of El Niño years during recent decades is rather high compared to the long-term statistics that are computed from about a thousand years of Nilometers data; however similar levels of activity have been observed during the first millennium.

The natural variability in the annual flow of the Nile is significantly regulated by the El Niño–Southern Oscillation (ENSO). In Chapter 4, several sources of information are combined, including ENSO, rainfall over Ethiopia, and the recent history of river flow in the Nile, in order to obtain accurate forecasts of the Nile flood at Aswan. The Bayesian theorem is used in developing the discriminant forecasting algorithm. Conditional categoric probabilities are used to describe the flood forecasts, and a synoptic index is defined to measure the forecasts' skill. The results presented show that ENSO information is the only valuable predictor for the long-range forecasts (lead time longer than the hydrological response timescale, which is 2–3 months in this study). However, the incorporation of the rainfall and river flow information in addition to the ENSO information significantly improves the quality of the medium-range forecasts (lead time shorter than the hydrological response timescale).

The human population living in the Nile basin countries is projected to double by 2050, approaching one billion. The increase in water demand associated with this burgeoning population will put significant stress on the available water resources. Potential changes in the flow of the Nile River as a result of climate change may further strain this critical situation. In Chapter 5, we present empirical evidence from observations and consistent projections from climate model simulations suggesting that the standard deviation describing interannual variability will increase by 50% in the twenty-first century compared to the twentieth century. We attribute the relatively large change in interannual variability of the Nile flow to projected increases in future occurrences of El Niño and La Niña events and to observed teleconnection between the El Niño–Southern Oscillation and Nile River flow. Adequacy of current water storage capacity and plans for additional storage capacity in the basin will need to be re-evaluated given the projected enhancement of interannual variability in the future flow of the Nile river.

As shown in Chapter 2, El Niño–Southern Oscillation (ENSO) explains about 25 % of the interannual variability in the Nile flow. In Chapter 6, we identify a region in the southern Indian Ocean, with a similarly strong teleconnection to the Nile flow. SST in the region (50–80° E and 25–35° S) explains 28% of the interannual variability in the flow of the Nile River and, when combined with the ENSO index, the explained variability of the flow of the Nile River increases to 44%. In addition, during those years with anomalous SST conditions in both oceans, this study estimates that indices of the SSTs in the Pacific and Indian oceans can collectively explain up to 84 % of the interannual variability in the flow of the Nile. Building on these findings, we use the classical Bayesian theorem to develop a new hybrid forecasting algorithm that predicts the Nile flow based on global model predictions of indices of the SST in the eastern Pacific and southern Indian oceans.

A significant fraction of the interannual variability in the Nile River flow is shaped by El Niño Southern Oscillation (ENSO). In Chapter 7, we investigate a similar role for the Indian Ocean (IO) SST in shaping the inter-annual variability of the Nile River flow.

Using observations of global SST distribution and river flow in addition to atmospheric general circulation model sensitivity experiments, we show that North and Middle IOSSTs play a significant intermediate role in the teleconnection between ENSO and the Nile flow. Applying partial coherency analyses, we demonstrate that the connection between North and Middle IO SSTs and Nile flow is strongly coupled to ENSO. During El Niño events, SST in the North and Middle IO increases in response to the warming in the Tropical Eastern Pacific Ocean and forces a Gill-type circulation with enhanced westerly low-level flow over East Africa and the Western IO. This anomalous low-level flow enhances the low-level flux of air and moisture away from the Upper Blue Nile (UBN) basin resulting in reduction of rainfall and river flow. SSTs in the

South IO also play a significant role in shaping the variability of the Nile flow that is independent from ENSO. A warming over the South IO, generates a cyclonic flow in the boundary layer, which reduces the cross-equatorial meridional transport of air and moisture towards the UBN basin, favoring a reduction in rainfall and river flow. This independence between the roles of ENSO and South IO SSTs allows for development of new combined indices of SSTs to explain the inter-annual variability of the Nile flow. The proposed teleconnections have important implications regarding mechanisms that shape the regional impacts of climate change over the Nile basin.

The Blue Nile originates from Lake Tana in the Ethiopian Highlands and contributes about 60–69% of the main Nile discharge. Previous studies investigated the relationship of SST in the Pacific Ocean (Niño 3.4 region) to occurrence of meteorological and hydrological droughts in the Nile Basin. In Chapter 8, we focus on the dependence of occurrence of droughts and floods in the upper catchment of the Blue Nile on the timing of El Niño and La Niña events. Different events start at different times of the year and follow each other, exhibiting different patterns and sequences. Here we study the impact of these timing and temporal patterns on the Blue Nile droughts and floods. The comparison between the discharge measurements (1965–2012) at the outlet of the upper catchment of the Blue Nile and the El Niño index shows that when an El Niño event is followed by a La Niña event, there is a 67% chance for occurrence of an extreme flood. Furthermore, we also found that 83% of El Niño events starting in April–June resulted in droughts in the upper catchment of the Blue Nile. Observations as well as global model forecasts of SST during this season could be used in seasonal forecasting of the Blue Nile flow.

In Chapter 9, we successfully simulate the observed statistical relationship between ENSO and the rainfall regime of the upper Blue Nile using the tropical-band version of the regional climate model RegCM4 (or Reg-TB). An ensemble of nine simulations for the 28-year period 1982–2009 is completed to investigate the role of ENSO in modulating rainfall over the upper Blue Nile catchment. Reg-TB shows a good skill in simulating the climatology of temperature, outgoing long-wave radiation patterns as well as related atmospheric circulation features during the summer season (i.e. the rainy season over the Blue Nile catchment). The model also succeeds in reproducing the observed negative correlation between Pacific SST and rainfall anomalies over the Blue Nile catchment, and in particular the association of droughts over the Blue Nile with El Niño events that start in April–June. We thus propose that observations and model forecasts of Pacific SST during this season could be used in seasonal forecasting of summer rainfall over the upper Blue Nile region.

The Chapters of this book offer detailed analysis of how conditions in the global oceans impact the hydrology of the Nile, and how these teleconnections can be used to improve seasonal forecasting of the Nile floods. The same teleconnections are also used to explain how climate change may impact the hydrology and climate of the Nile basin. All those interested in understanding past or future variability in the Nile floods will find relevant lessons in the following pages of the book.

Chapter 2—El Niño and the Natural Variability in the Flow of the Nile River

Elfatih A. B. Eltahir

Recent studies [e.g., Ropelewski and Halpert, 1987; Simpson et al., 1993] indicate that oscillations in the state of the Pacific region's ocean-atmosphere system are related to inter-annual fluctuations of rainfall and river flow in several regions of the world. These oscillations are known as El Niño-Southern Oscillations (ENSO). The early paper of Bliss [1925] lists the Nile flood as one of 10 geophysical variables that are related to a Southern Oscillation index. The recent study of Quinn [1992] explores the use of the historical record of maximum Nile flood to extend the records of the Southern Oscillation index.

In a recent paper, Moss et al. [1994] use the Southern Oscillation index as a predictor of the probability of low streamflows in New Zealand. My research group at the Massachusetts Institute of Technology published a study in 1996 testing the hypothesis that the natural variability in annual flow of the Nile River at Aswan is related to ENSO and that such information can be used for improving the predictability of the Nile flood.

We presented a simple hypothesis to explain some of the statistical characteristics observed in the time series of Nile flow. In particular, we proposed that the relation between ENSO and the Nile flood may explain the Hurst phenomenon: the mean of the annual flow process in the Nile River varies in the time following ENSO, resulting in a non-stationary process, and causing the Hurst phenomenon (where the term nonstationary is used to indicate that the mean of the process is a variable function of absolute time). The data, methods and findings are detailed in the remainder of this chapter.

Data

The data used in this analysis consist of annual flow in the Nile River at Aswan for the century that extends between the years 1872 and 1972. The monthly distribution of the Nile flow is described by Figure 2.1, which shows the long-term average flow for each month. The annual flow of the Nile (the Nile flood) is defined as the cumulative flow measured from June of any year to May of the following year. A typical monthly flow in the Nile for any year closely follows a seasonal cycle, with one peak and a long recession. Hence most of the natural variability in the Nile flow is captured by the annual flow. The measured flow of the Nile has been naturalized by removing anthropogenic trends, but the findings of this paper are not sensitive to this correction. This data set on the Nile flow has been supplied by the Ministry of Water Resources of Egypt.

The index of ENSO used in this study is a homogenized monthly series of the mean sea surface temperature (SST) anomaly averaged over the regions 6°– 2°N, 170°–90°W; 2°N – 60S, 180°– 90°W; and 6°– 10°S, 150°– 110°W during the same time period, 1872-1972. This index of ENSO was published by Wright [1989]. Predictions of SST in

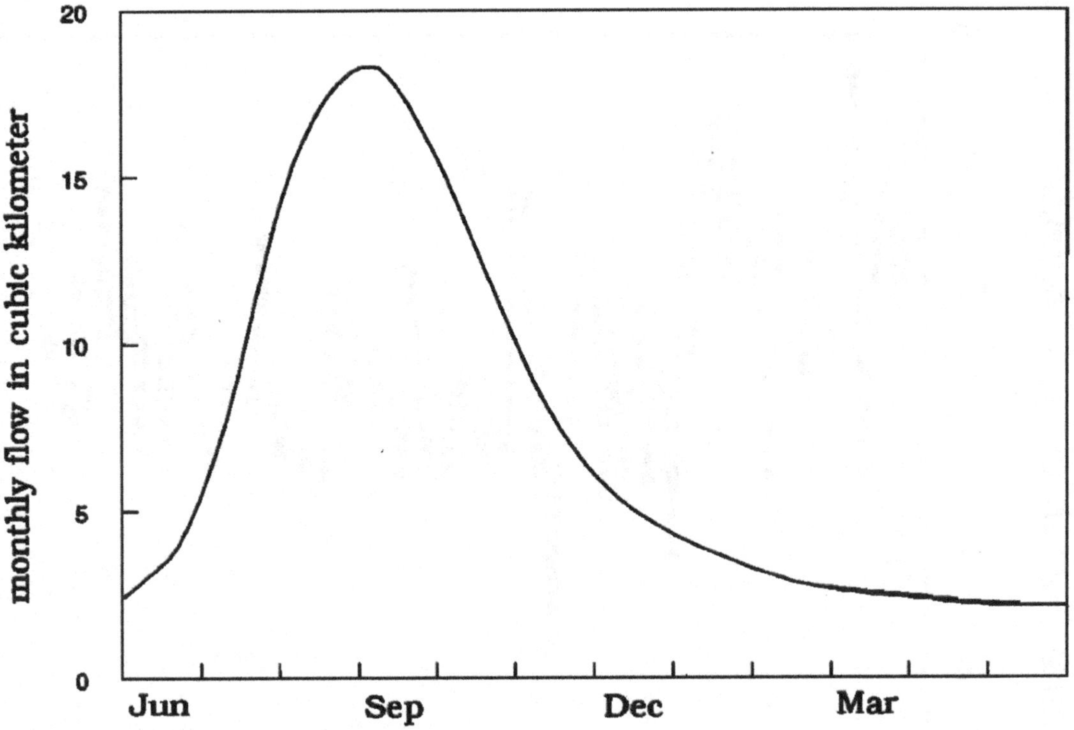

Figure 2.1: Seasonal cycle of the Nile flow at Aswan

the Pacific Ocean were provided by Stephen E. Zebiak of Lamont-Doherty Geological Observatory, Columbia University.

Relation Between ENSO and the Nile Flood

The natural series of annual flow of the Nile at Aswan is shown in Figure 2.2. The years with El Niño events are defined according to the classification of Rasmusson and Carpenter [1983]) and are marked in Figure 2.2 with solid circles. From Figure 2.2 we may conclude that ENSO events are associated with those years when the flow of the Nile is low compared to the long-term average. This observation prompted a more careful

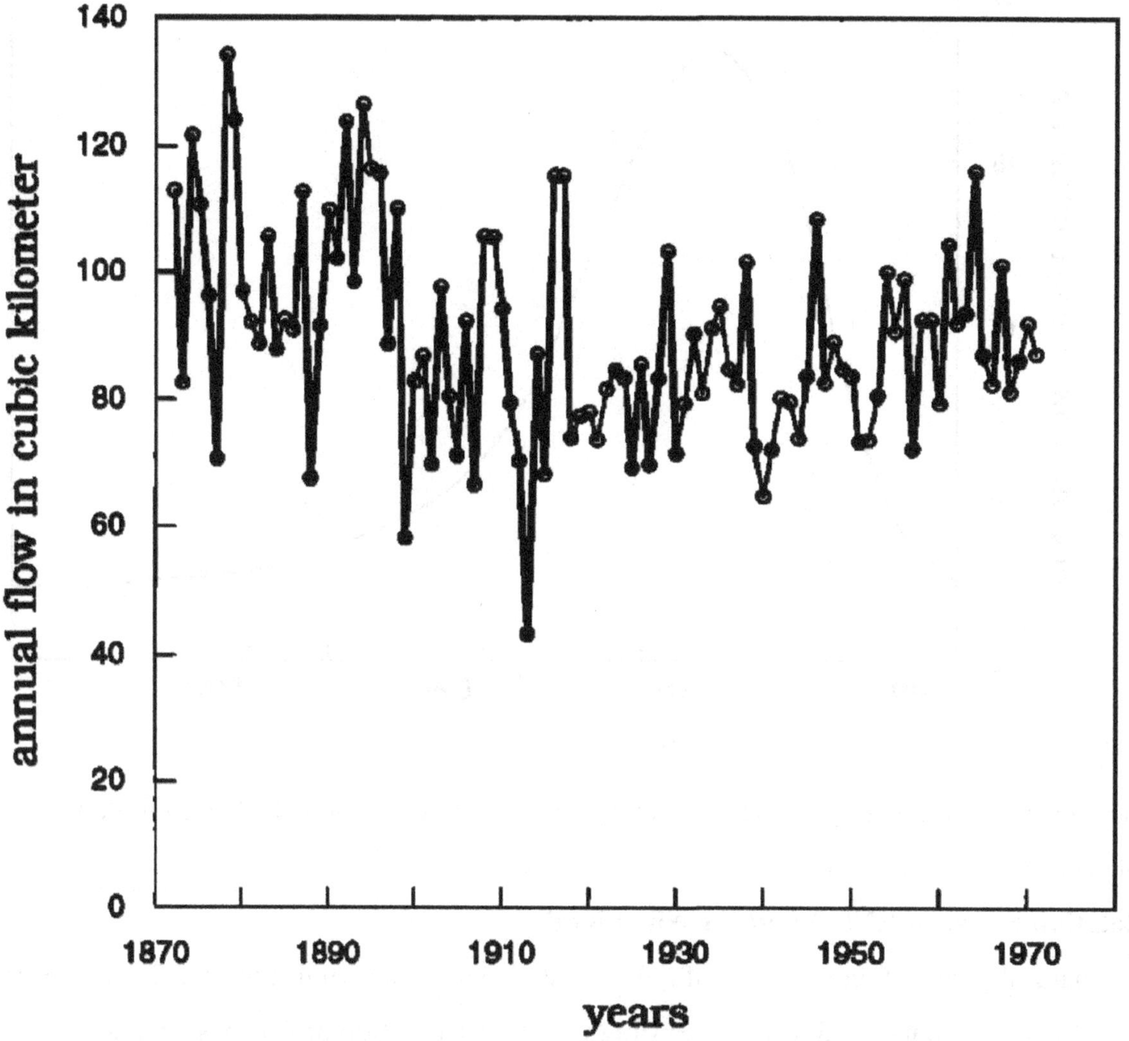

Figure 2.2: Annual fluctuations of the "natural" flow in the Nile at Aswan for the years 1872-1972. El Niño events are marked with solid circles.

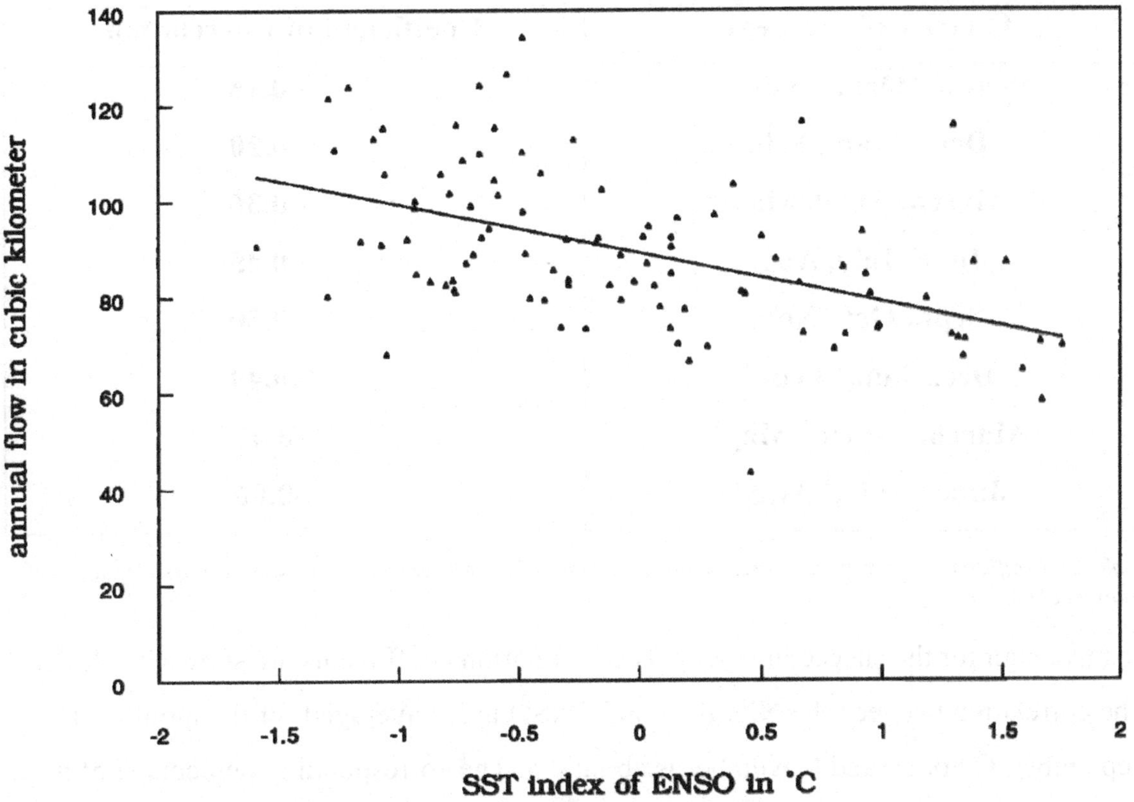

Figure 2.3: Scatter plot of the data showing the relation between annual Nile flow and the index of ENSO for the months of Sept, Oct, and Nov, for the years 1872-1972. The coefficient of correlation is -0.5, The straight line is the regression line.

look at the correlations of ENSO indices and the Nile flow. The index of ENSO is averaged over four quarters of the year: December to February, March to May, June to August, and September to November. Then the correlations between the annual flow of the Nile (the flow from June to May) and the average index of ENSO for each of these quarters are computed. The different quarters are denoted by negative sign when the correlation is computed between the Nile flood and SST in the preceding year, and a

Quarter of the Year	Coefficient of Correlation
Sept.,$^{-}$ Oct.,$^{-}$ Nov.$^{-}$	**- 0.15**
Dec.,$^{-}$ Jan., Feb.	**- 0.20**
March, April, May	**- 0.36**
June, July, Aug.	**- 0.45**
Sept., Oct., Nov.	**- 0.50**
Dec., Jan.,$^{+}$ Feb.,$^{+}$	**- 0.49**
March,$^{+}$ April,$^{+}$ May^{+}	**-0.42**
June,$^{+}$ July,$^{+}$ Aug.$^{+}$	**-0.06**

Table 2.1: Coefficients of correlation between the annual flow of the Nile and Index of ENSO averaged for different quarters of the year

positive sign for the succeeding year. The correlation coefficients are shown Table 2.1. The correlation between the Nile flow and ENSO index averaged for the months of September, October, and November is about 0.5. The corresponding percentage of the variance in annual flow explained by the ENSO index is 25%. The scatter plot for this relation is shown in Figure 2.3.

An explanation is provided for the relation between ENSO and the annual flow of the Nile based on observed correlations in the global distribution of sea level pressure. In particular, we note the observed positive correlation between the annual mean sea level pressure at the sources of the Nile, in the Ethiopian Plateau, and the annual pressure at Darwin, Australia [see Trenberth and Shea, 1987]. A positive anomaly in the annual surface pressure at Darwin is an indicator of the onset of an ENSO event. But according

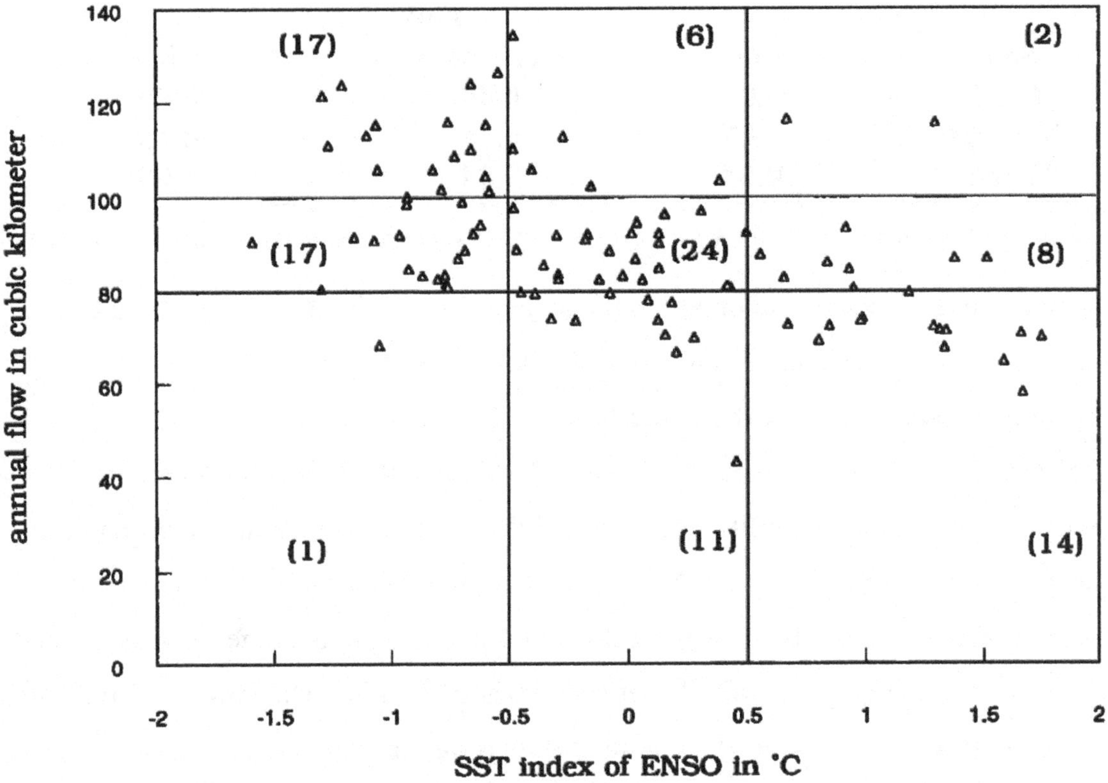

Figure 2.4: Categories of the Nile flood and ENSO index. The number of data points in each group is also shown.

to the observations, Darwin pressure is positively correlated with pressure anomalies over the Ethiopian Plateau. The laws of atmospheric dynamics require that a positive anomaly in annual sea level pressure over the Ethiopian Plateau is associated with decreased convergence of atmospheric moisture and therefore less rainfall. The observational study of Janowiak [1988] confirms that ENSO events are associated with drier than normal levels of summer rainfall in a region that includes the Ethiopian Plateau. This reduction

	Flow		
SST	**Low**	**Average**	**High**
Cold	**0.02**	**0.49**	**0.49**
Normal	**0.27**	**0.58**	**0.15**
Warm	**0.58**	**0.34**	**0.08**

Table 2.2: Conditional probability of the Nile flood given the SST Index of ENSO based on observations of 1872-1972.

in rainfall causes a negative anomaly in the annual flow of the Blue Nile. The latter contributes most of the inter-annual variability in the Nile flow.

ENSO and Predictability of the Nile Flood

The observed correlation between ENSO and the flow in the Nile has important implications on the predictability of the annual flow in the river. A simple probability table is developed for use in predictions of the Nile flood. This table is based on the classification that is described in Figure 2.4 and Table 2.2. An average flood is defined as the annual flow that has a magnitude between 80 and 100 km^3. The long-term average for the Nile flood is about 90 km^3. If the annual flow is larger than 100 km^3, it is classified as high. If the same flow is smaller than 80 km^3, then it is classified as low. The data on SST are classified similarly into cold, normal, and warm conditions. The boundaries between these three conditions are temperatures of -0.5°C and 0.5°C, respectively.

The choices of the boundaries between the different conditions for SST as well as the Nile flood are subjective. Indeed, the same exercise can be performed using any alternative reasonable choices.

Table 2.2 describes the prediction probabilities for the Nile flood. This table consists of the conditional probability of having a

Nile flood in a certain category (low, average, or high) given a certain observed SST condition (cold, normal, or warm). For example, the probability of having a low flood given a cold SST condition is 2%; and the probability of having a high flood given a warm SST condition is 8%. In order to appreciate the information content in relation to ENSO and the Nile flood, we note that in absence of any information about SST, the probability of having a low flood in any year is 26%, and the corresponding probability for having a high flood is 25%. These estimates are based solely on the past records of Nile floods. Hence knowledge of SST conditions could potentially be very useful in prediction of the Nile flood.

Recent studies presented optimistic conclusions about the predictability of ENSO events, suggesting that these events could be accurately predicted one or two years ahead using a coupled model of the ocean-atmosphere system [see Cane et al., 1986]. In this study, we utilize predictions of SST in the Pacific Ocean, which are obtained using a coupled ocean-atmosphere model, for the purpose of making predictions about the Nile flow. The predictions of SST are obtained using an improved version of the model described by Cane et al. [1986].

These predictions are issued 9 months in advance and were supplied to us by Stephen E. Zebiak of Lamont-Doherty Geological Observatory of Columbia University. Table 2.3 shows predictions of the SST index of ENSO for the years 1973-1989. These predictions are used to make similar predictions for the Nile flood in the same years, made as early as the February that precedes the Nile flood of interest. The observed floods for the same years are shown in Table 2.2 for verification purposes.

Predictions of Flood (Probability)							
	Predicted SST		%			Observed Flood	
Year	**°C**	**Category**	**Low**	**Average**	**High**	**km³**	**Category**
1973	-0.6	cold	2	49	49	83	average
1974	-0.6	cold	2	49	49	90	average
1975	-0.1	normal	27	58	15	102	high
1976	0.9	warm	58	34	8	80	average
1977	1.3	warm	58	34	8	92	average
1978	-0.3	normal	27	58	15	86	average
1979	0.0	normal	27	58	15	69	low
1980	-0.3	normal	27	58	15	82	average
1981	0.0	normal	27	58	15	85	average
1982	2.5	warm	58	34	8	69	low
1983	2.1	warm	58	34	8	72	low
1984	1.7	warm	58	34	8	53	low
1985	0.5	normal	27	58	15	82	average
1986	1.0	warm	58	34	8	73	low
1987	1.9	warm	58	34	8	68	low
1988	0.2	normal	27	58	15	115	high
1989	0.4	normal	27	58	15	81	average
The flood predictions are described by probabilities of having a certain flood level							

Table 2.3: Predictions of SST and predications and observations of the Nile flood 1973-1989.

In 12 of the 17 years the prediction of the most probable flood matches the observations. For example, the prediction for the most probable flood for 1984 is a low flood level that matches the observation for that year. Since the autocorrelation coefficient at lag 1 in the Nile flood series is insignificant (-0.22), there is very little persistence that can be used for forecasting purposes. Hence in absence of any information about SST, the prediction of the most probable flood for any of the years will be the average flood. According to the observations for 1973-1989, this prediction matches the observations in 9 out of the 17 years. From this comparison, we conclude that the use of ENSO predictions could improve the Nile flood predictions significantly.

The value of the probability table for prediction purposes can be further appreciated by considering the extreme flood conditions. A prediction of a warm SST translates into a very low probability (8%) of having a high flood. On the other hand, a prediction of a cold SST translates into an even smaller probability (2%) of having a low flood. It is encouraging none of the years between 1973 and 1989 had cold SST/low flood or a warm SST/high flood conditions. For the purpose of water resources management, the ability to (almost) rule out the possibility of occurrence of a high flood (or a low flood) 9 months ahead, could be of great economic value.

ENSO and the Hurst Phenomenon

Hurst [1951] observed that the time series of annual flow in the Nile River exhibits statistical characteristics that are not compatible with a series of purely random fluctuations: an observation that has come to be known as the Hurst phenomenon. This

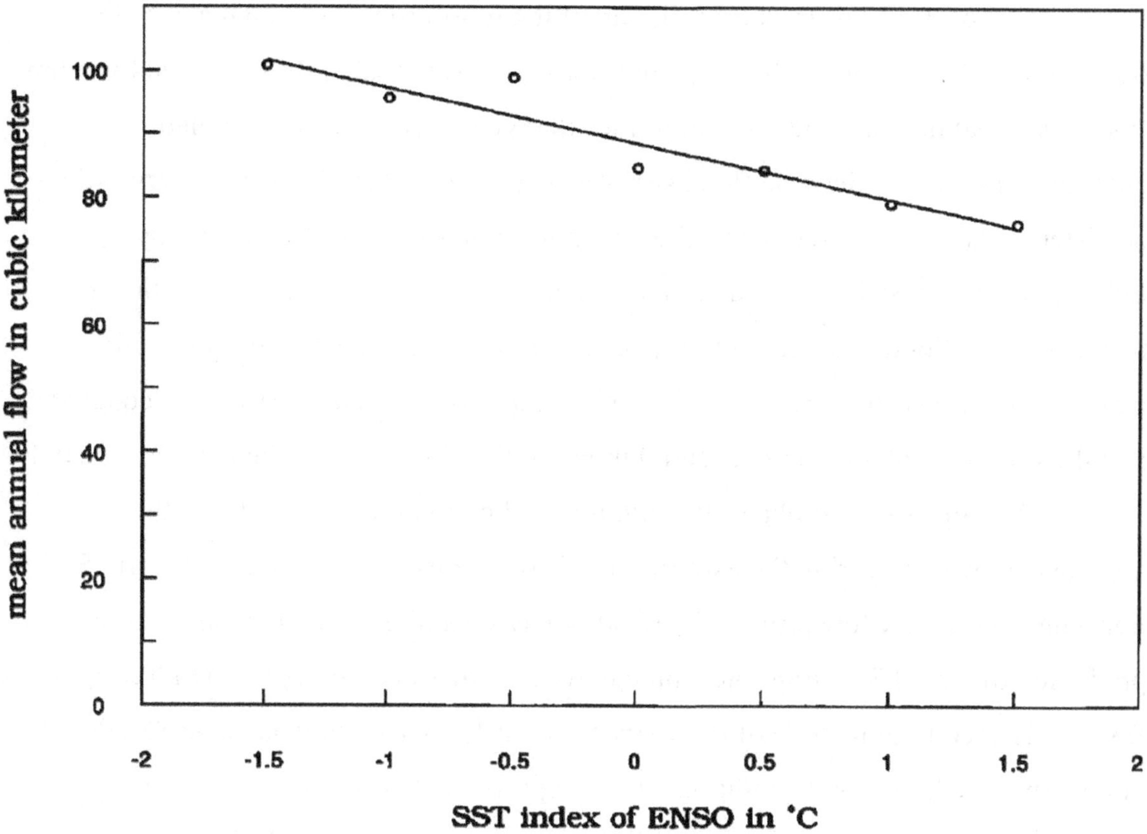

Figure 2.5: The relation between ENSO and the varying mean of the annual flow in the Nile River. The annual flow of the Nile is averaged over a small window of ENSO indices, each with a width of 0.5°C centered around (-1.5°, -1.0°, -0.5°, 0.0°, 0.5°, and 1.5°C). The number of points within each of these windows are 4, 19, 25, 21, 11, 9, and 11, respectively.

observation is most relevant to the hydrologic design of inter-annual storage reservoirs. For example, the size of a reservoir on the Nile River, which on average fails once every n years in supplying exactly the long-term mean flow, is roughly proportional to $n^{o.72}$. In contrast, for any purely random river flow the same reservoir size is proportional to $n^{o.5}$.

Hence for any specific design criteria, a reservoir on the Nile has to be of a relatively large size.

Several studies attempted to explain the Hurst phenomenon based on statistical arguments and models. Bras and Rodriguez-Iturbe [1985] summarize the different lines of thought in explaining the Hurst phenomenon as a result of (1) a transitory behavior due to the short records of observed geophysical variables [see Salas et al., 1979, Figure 2.5], (2) nonstationarity of the underlying mean of the process [see Klemes, 1974; Potter, 1976; Boes and Salas, 1978; Bhattacharya et al., 1983], and (3) a stationary process with very large memory.

We hypothesize a physically-based explanation for the Hurst phenomenon in the Nile flow. The above analysis indicates that the occurrence of ENSO events excites similar oscillations in the tropical climate, which are then teleconnected to the Nile flow through the rainfall-producing mechanisms at the sources of the Nile. The relation of the ENSO index and the Nile flood suggests that the Nile flood responds to natural variability in SST of the Pacific Ocean. This variability occurs at several timescales ranging from annual to decadal and even longer. As a result, variability in SST of the Pacific Ocean causes significant nonstationarity in the mean of the annual flow process in the Nile River. Natural climatic factors (other than ENSO) dictate the variability of the Nile flow around the varying mean of the process.

The relation between ENSO index and the nonstationary mean of the annual flow in the Nile is estimated. This relation is shown in Figure 2.5. The annual flow of the Nile is averaged over a small window of ENSO indices, each with a width of 0.5°C centered around - 1.5°, - 1.0°, -0.5°, 0.0°, 0.5°, 1.0°, and 1.5°C. The number of points within each of these windows are 4, 19, 25, 21, 11, 9, and 11, respectively. Each group of points is treated as a sample from a different population. Although some of these samples have only few data points, the nonstationary mean of the Nile flood is clearly related to ENSO index. The resulting relation is described by mean of the annual flow of the Nile

= 88.5-8.7 (ENSO index) (1).

The coefficient of correlation in this relation is -0.9. The varying mean of the annual flow in the Nile River is computed in Figure 2.6 using (1) and the series of observed ENSO indices. Figure 2.6 shows the varying mean for the years 1872-1972. These random oscillations represent the contribution of the ENSO index to the natural variability in the flow of the Nile River.

Based on this analysis, we propose the following hypothesis: the mean of the annual flow process in the Nile River varies in time following ENSO, resulting in a non-stationary process, and causing the Hurst phenomenon. The statistical characteristics of nonstationary processes have been studied extensively, and their ability to reproduce the Hurst phenomenon has been proven [see Potter, 1976; Boes and Salas, 1978]. In order to exclude effects due to the transitory behavior of the process, the hypothesis presented in this paper can only be rigorously tested when long time series (- 10^3 data points) of the annual flow in the Nile River becomes available.

Our hypothesis about the Hurst phenomenon is similar to the argument offered by Potter [1976]. He analyzed precipitation records from North America and pointed to the potential role of oscillations in the state of the global ocean-atmosphere-land system in

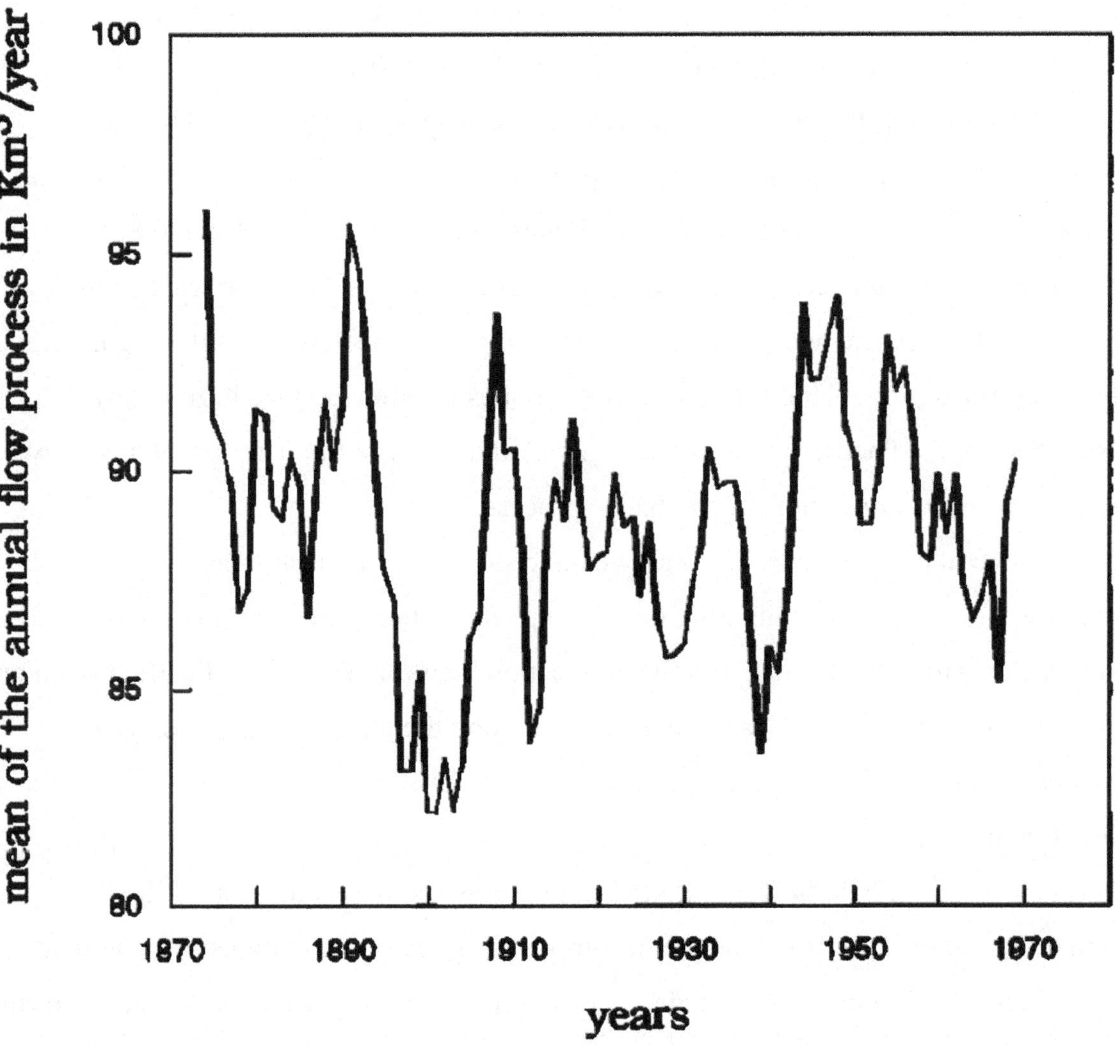

Figure 2.6: The varying mean of the annual flow process estimated from observed ENSO indices and using (1).

providing a physical explanation for the Hurst phenomenon. However, this analysis has focused on a specific global oscillation—the ENSO—and the hypothesis presented is relevant to a different natural variable—the Nile flood.

In summary, the hypothesis described in this section is motivated by the following: The analysis presented in this paper confirms the existence of a significant relation between the mean of the Nile flood and SST of the Pacific Ocean (see Figure 2.5). The SST of the Pacific Ocean exhibits oscillations at several timescales, ranging from annual to decadal and even longer [see Wang and Ropelewski, 1995, Figure 2.5]. Hence the mean of the Nile flood process oscillates in time, too (see Figure 2.6). As a result, SST of the Pacific Ocean (ENSO index) and the Nile flood represent a weakly coupled bivariate nonstationary stochastic process.

Several of the studies listed above have shown that nonstationary stochastic processes are capable of simulating the Hurst phenomenon. The combination of these three factors suggests that offering the connection between SST of the Pacific Ocean and the Nile flood as a potential candidate for explaining the Hurst phenomenon is a reasonable hypothesis.

Conclusions

The analysis of the Nile flow and ENSO index suggests that natural variability in the annual flow of the Nile River can be decomposed into two components: a mean that varies in time following ENSO, and a random fluctuation that occurs around the varying mean due to climatic factors other than ENSO. The nonstationary mean and the random fluctuation explain 25% and 75% of the observed natural variability, respectively. The relation between ENSO and the Nile flood has important implications on two important problems: predictability of the Nile flood, and explanation of observed statistical characteristics of the Nile flood series (the Hurst phenomenon).

The fact that ENSO events can be predicted with reasonable accuracy at a lead time of about one year suggests that the correlation between ENSO and the Nile flood should be used to improve the predictability of the annual flow in the Nile River. The finding that the Nile flood series can be decomposed into a varying nonstationary mean and a random fluctuation favors the line of thought that the Hurst phenomenon may be due to nonstationarity in the mean of the process.

For this reason, we presented in our 1996 paper the hypothesis that ENSO is the causal factor for the Hurst phenomenon in the annual flow of the Nile River, a hypothesis that has proven out in the years since. We proposed a simple procedure for prediction of the Nile flood and illustrated by application for the years 1973-1989. We computed the conditional probabilities of having a low, average, or high flood in the Nile River given the SST conditions in the Pacific Ocean. The new procedure was designed to utilize model predictions of SST in the Pacific Ocean. In the absence of this procedure, the best practical prediction of the Nile flood would be the mean flow (-90 km^3). The analysis presented in this paper suggested that our proposed procedure improves on this prediction. Such an improvement in the Nile flood prediction should have a positive impact on water resources management in Egypt, particularly in operation of the High Aswan Dam.

Future research will focus on further development of the flood prediction procedure. The emphasis will be in combining SST measurements and predictions with other inputs, such as hydrological measurements of rainfall and streamflow to improve the prediction of the Nile flood. Such efforts are outlined in the following chapter.

A version of this chapter first appeared as

Eltahir, E., 1996. El Niño and the Natural Variability in the Flow of the Nile River. Water Resources Research, 32(1), pp.131-137.

Chapter 3—Nilometers, El Niño, and Climate Variability

Elfatih A. B. Eltahir
Guiling Wang

The Nile water was of critical importance to the ancient civilization of Egypt, and remains so to the society built in the region today. Nilometers have been used for gauging the level of water in the Nile river for more than five millennia. Written records describing some of these measurements are available for most of the last fourteen centuries and they perhaps represent the longest written records for any hydrological phenomenon. Quinn [1992] analyzed these records carefully, with the objective of extending the recorded history of the El Niño-Southern Oscillation (ENSO) phenomenon. He developed an index that is based on the fluctuations in the Nile water level. Diaz and Pulwarty [1992] compared this Nile index to other ENSO indices and confirmed that the compilation of the Nile flood intensity record by Quinn represents a valuable contribution to ENSO studies. This recorded history of the Nile flood has been used in studying the long-term natural variability of the El Niño phenomenon [Anderson, 1992]. In the research outlined in this chapter, we extended the Nile flood record up to 1997 and used it to study the statistical significance of the then-recent trend in this phenomenon.

The observed increase in the frequency of El Niño years during the last few decades may be partly caused by the evident increase in the concentration of greenhouse gases [Trenberth and Hoar, 1996]. This hypothesis is consistent with the results of a the 1996 study by Meehl and Washington that simulated the response of a numerical model describing the coupled global ocean-atmosphere system to the increase in the concentration of atmospheric carbon dioxide. We determined that this phenomenon deserves further investigation.

In particular, we needed to evaluate accurately the statistical significance of the recent trend in El Niño occurrences. Considerable differences existed between the conclusions of the different studies on the statistical significance of the observed change in the frequency and duration of El Niño events during the 1980s and 1990s [Trenberth and Hoar, 1997; Harrison and Larkin, 1997; Rajagopalan et al, 1997]. Estimates of the corresponding return periods range from a few hundreds to a few thousands of years. These estimates were based on statistical analysis of relatively short records of El Niño indices. There was a need for a consistent analysis on a long-term record of the El Niño years in order to determine the statistical significance of the trend in this phenomenon.

In this research, which we undertook in the late 1990s and published in 1999, we used the history of observations of water level in the Nile river to test the statistical significance of the trend in the El Niño phenomenon. The well-established association between El Niño and the Nile river flow was supported by extensive observational evidence [Bliss, 1925; Quinn, 1992; Aramarasekera et al, 1997; Chapters 2 and 4 of this book]. It is worth noting that the coherency between the ENSO index and the Nile flood has a distinguished peak at the time scale of 4 to 5 years, which is close to the ENSO time scale (see Figure 2.3c of this book). As we described in Chapter 2, warming in the equatorial eastern Pacific is associated with rising motion over the eastern and central Pacific ocean and sinking motion over most of the remainder of the tropics, including the sources of the Blue Nile over the Ethiopian Plateau [also see Aramarasekera et al, 1997]. The East African monsoon is the main rainfall-producing mechanism over the Blue Nile catchment and provides the teleconnection between El Niño and the Blue Nile flow. During the flood season, from July to October, most of the water in the main Nile is supplied by the Blue Nile. Thus, a warm Sea Surface Temperature (SST) in the Pacific ocean is associated with anomalous sinking motion over the Indian ocean and surrounding land regions, less rainfall over Ethiopia (see more in Chapter 4), and drought conditions in the river flow of the Blue Nile and that of the main Nile [Aramarasekerae t al, 1997; also Chapter 2 of this book].

Naturalization of the Nile Data

The interannual fluctuations in the Nile water level during the summer months should reflect the natural variability in the East African monsoon and the El Niño phenomenon. However, during the 20th century the extent of human impact on the natural flow of the Nile river has reached a point where direct measurements of the river flow may no longer be indicative of the natural variations in the monsoon system. In particular, the extraction

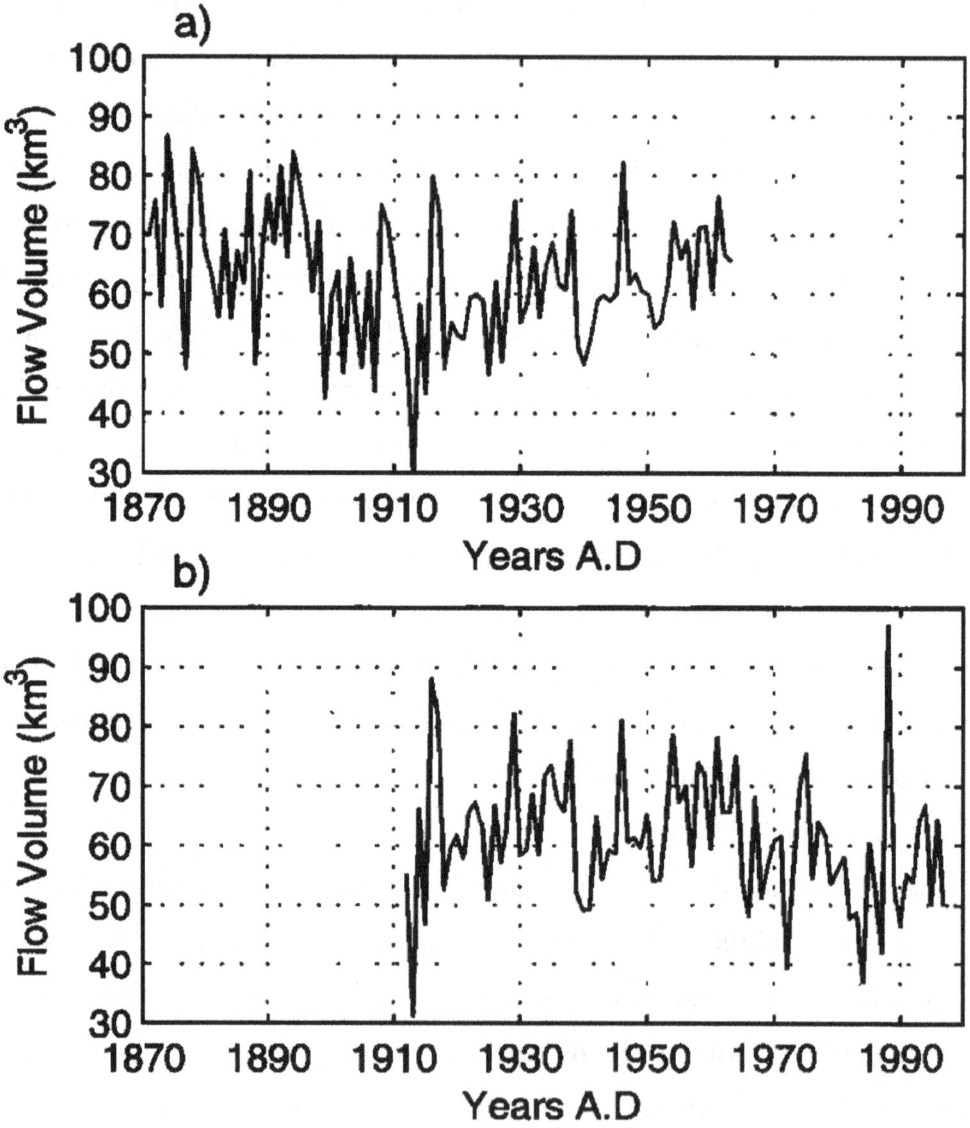

Figure 3.1: Naturalized flood volume in summer (July to October) at (a) Aswan for the years 1871 – 1964; and (b) Dongola for the years 1912 – 1997

of water from the Blue Nile for irrigation purposes in Sudan, and the construction of the High Aswan dam had noticeable effects on the observed fluctuations of river flow.

Careful consideration was given in this study to the elimination of the effects of the two anthropogenic factors mentioned above. First, we used two complementary observational records from two nearby stations—Aswan and Dongola—to characterize the recent history of the Nile flow. Dongola is located on the Nile river at about 600 kilometers upstream from Aswan. The flow records at Dongola are useful for characterizing the natural flow of the Nile after the construction of the High Aswan dam.

Second, we estimated water extraction in Sudan. The extraction of water occurred after 1925 when Sennar dam was built to supply water for the Gezira irrigation scheme. The water extraction can be estimated by the change, with reference to the period before 1925, in the difference between the observed flow at Roseris station and Khartoum station for the period 1912-1997. Both stations are located on the Blue Nile. Roseris is located near the Sudan/Ethiopia border, upstream from where the extraction occurred. Khartoum is located at the confluence of the Blue Nile and the White Nile, downstream from where the extraction occurred. After estimating the water extraction, the Nile flood series at both Aswan station and Dongola station can then be naturalized by adding the water extraction estimates to the observed series.

Figure 3.1 shows the resulting naturalized series of the summer flood at Aswan and Dongola. These two series are highly correlated, and they collectively describe the history of the Nile river flow for the period 1871- 1997.

El Niño Index Based on Nilometer Records

The normalized fluctuations of the summer flood for Aswan and Dongola are presented in Figure 3.2 in comparison to the normalized fluctuations of the SST within the Pacific Ocean (a standard index of El Niño). The SST data used here is from the UKMO global data set. It has been averaged for the regions 6-2N, 170-90 W; 2N-6S, 180-90 W; and 6-10S, 150-110W (same as the area used by Wright, [1989]. The SST records have also been averaged from September through November, which (as we saw in Chapter 2) has the most significant correlation with the Nile flood. The coefficient of correlation between Nile flood and SST in the Pacific ocean is about -0.54. This coefficient is quite significant, especially if we compare it to the corresponding coefficient for the correlation between the SST index of El Niño and the Southern Oscillation Index (SOI) which is about - 0.78.

However, examination of Figure 3.2 reveals the non-stationary nature of the association between the Nile floods and ENSO. During the decades of the 1930s and 1940s, this correlation is less significant than in other periods. Hence, we emphasize that the analysis described in this report assumes that the correlation between the two phenomena remains always significant [Allan et al, 1998]. We analyzed the naturalized series of the Nile river flow for the last 127 years using the same criterion that has been developed by Quinn [1992]. This analysis estimates the order of El Niño in each year based on the departure of summer (July-October) flood volume at Aswan station from the 50-year moving average. Order 1 corresponds to a departure of somewhere between 5.5 km^3 to 10.9 km^3; order 2 corresponds to a departure between 10.9 km^3 and 16.4 km^3;

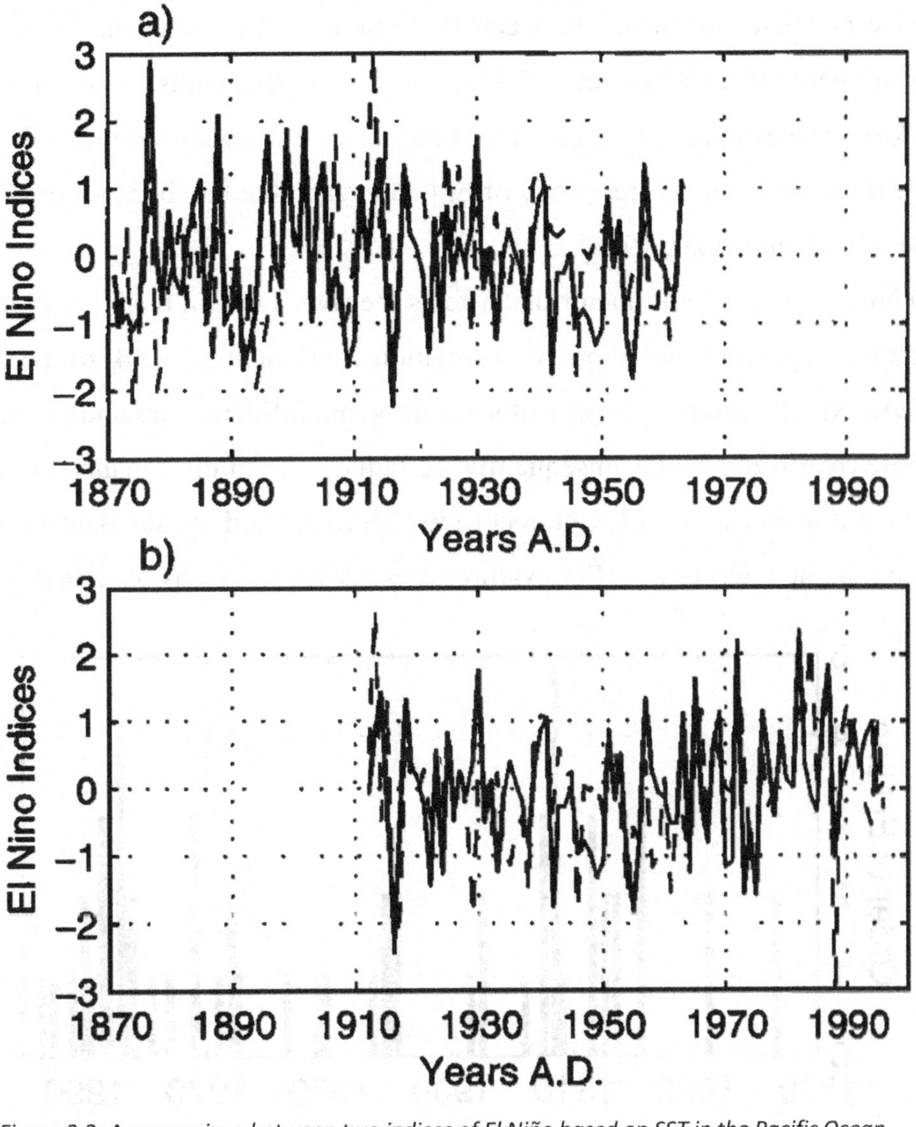

Figure 3.2: A comparison between two indices of El Niño based on SST in the Pacific Ocean (continuous line) and the summer flood of the Nile river (dashed line) at (a) Aswan; and (b) Dongola

order 3 corresponds to a departure between 16.4 km^3 and 21.8 km^3; order 4 corresponds to a departure between 21.8 km^3 and 27.3 km^3; order 5 corresponds to a departure larger than 27.3 km^3. The estimated El Niño order time series is shown in Figure 3.3, which reflects the recent increase in frequency of El Niño years that had been reported by other studies [Trenberth and Hoar, 1996].

El Niño is not the only factor that affects the river Flow in the Nile. In fact, it only explains about 30 percent (based on the correlation coefficient of -0.54) of the variability in the Nile flood. There must be some other factors that influence the rainfall generation over the Nile river watershed. Consequently, certain degree of discrepancy exists between the Nilometer-based El Niño record in Figure 3.3 and the standard El Niño record based on the Pacific SST. For example, a weak La Niña was observed in 1984,

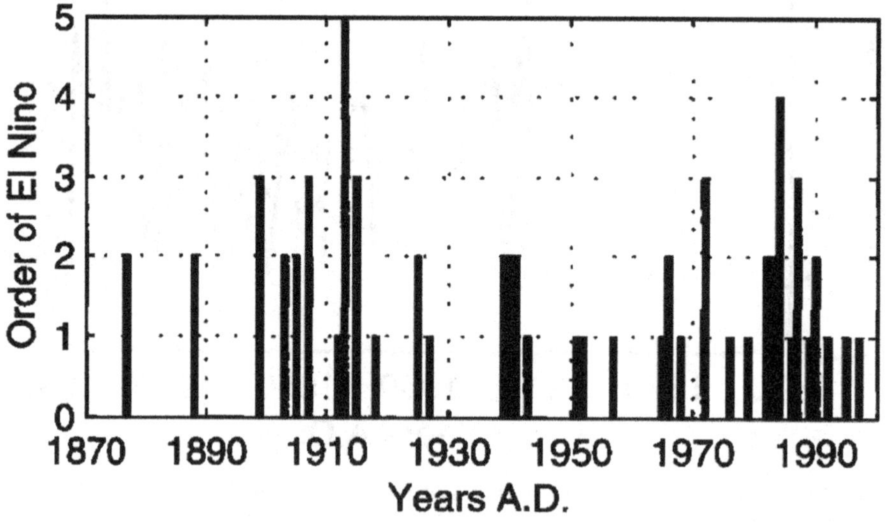

Figure 3.3: Order of El Niño years estimated based on the Nile flow series of Figure 3.1

while the Nile flood record signaled a strong El Niño in that year. Discrepancy in El Niño strength also exists in some El Niño years. However, in general, the Nilometer record still captures most of the variability of El Niño phenomena.

In order to study the low frequency variability of any phenomenon we need to consider records that are long enough to resolve those low frequencies. Otherwise, we have to assume that the conditions experienced in the short record are representative of longer periods. In order to avoid such assumptions, we used the recorded history of the Nile river flow as a long-term record of the El Niño phenomenon. By taking this approach, we gained additional information about the low frequency variability of this phenomenon at the expense of sacrificing some of the accuracy in our estimates of the frequency of El Niño years. In our analysis, the El Niño order by Quinn [1992] based on Nilometer records were used for the period 622-1522 A.D., and the El Niño order series of Figure 3.3 are used for the period 1871-1997 A.D.. The Nilometer records between 1522-1870 A.D. have not been included here since, for political reasons, those records include several gaps.

Results and Conclusions

Figure 3.4a describes consistent estimates of the frequency of El Niño years that are based on the relatively recent and short record of observations on the Nile river flow (Figure 3.3) as well as long-term Nilometers records. Any year for which the order of El Niño is greater than zero is identified as an El Niño year. The frequency of El Niño years was computed using a 21-year moving average. Figure 3.4b shows the return period of the El Niño years, derived empirically using the frequency record of Figure 3.4a.

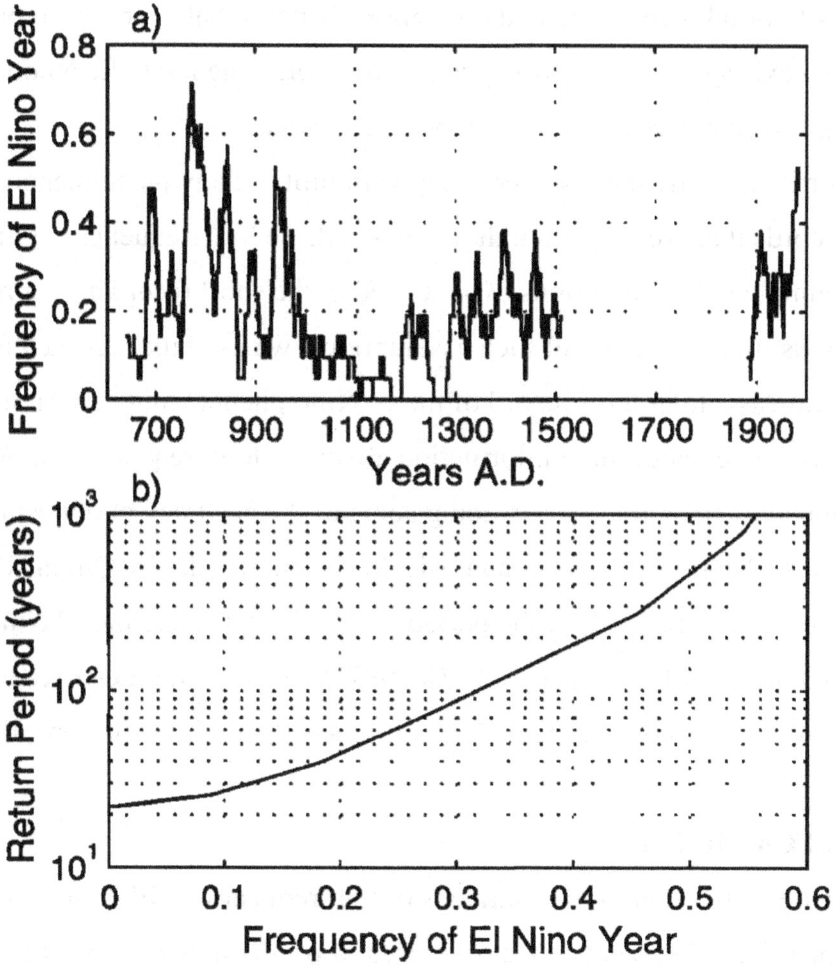

Figure 3.4a: Frequency of El Niño years based on the estimation of the El Niño order by Quinn (1992) for the period 622 – 1522 A.D., and based on the El Niño order series of Figure 3.3 for the period 1871 – 1997 A.D.

Figure 3.4b: The return period describing the frequency of El Niño years.

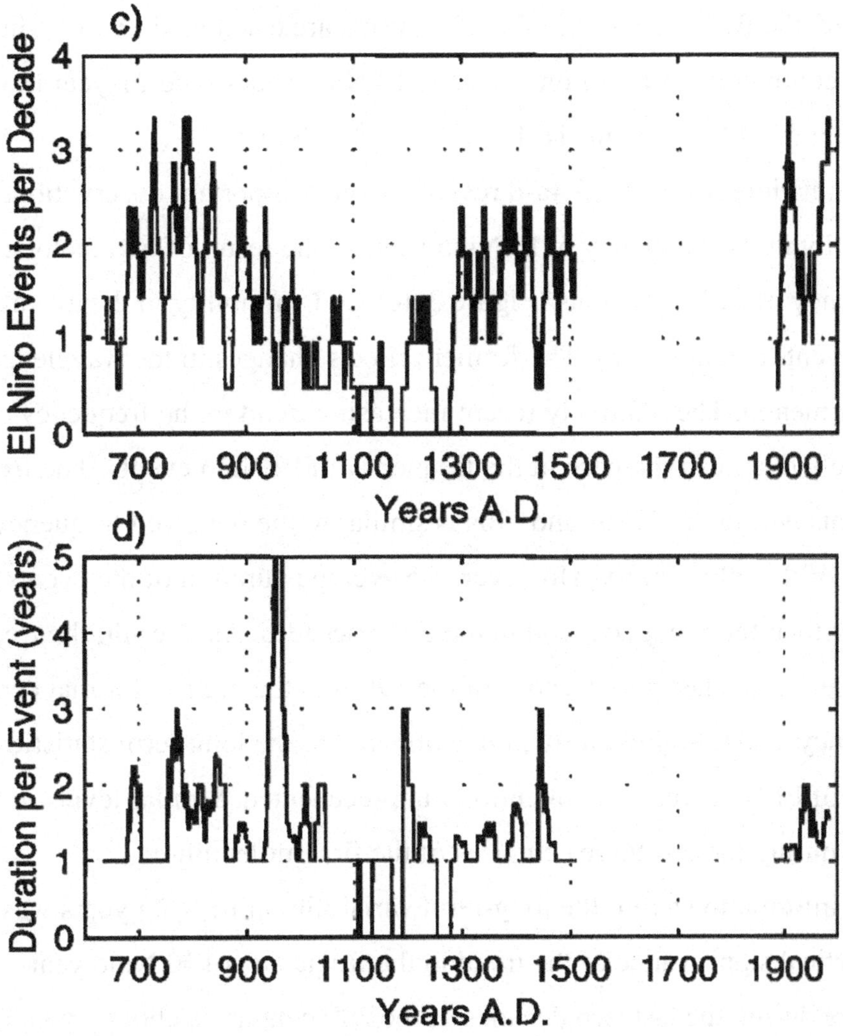

Figure 3.4c: Frequency of El Niño events.

Figure 4.4d: Average duration of El Niño events

In Figure 3.4c, the frequency of the El Niño events are calculated. An El Niño event is defined as a continuous series of one or more El Niño years. The 21-year moving average of the duration of El Niño events is shown in Figure 3.4d.

Examination of Figures 3.4a-d reveals several important observations. The long-term variability in the frequency of El Niño years is the result of two factors: variability in the frequency of El Niño events (Figure 3.4c), and variability in the average duration of El Niño events (Figure 3.4d). The former reflects changes in the wavelength of the ENSO phenomenon. The relatively recent increasing trend in the frequency of El Niño years is largely due to an increase in the frequency of El Niño events. The frequency of El Niño events within the 1980s and '90s is similar to the observed frequency in the first few decades of the 20th century. However, the average duration of the events in the latter half is longer than those experienced in the early decades. As a result, the frequency of El Niño years during the last two decades of the 1900s is the highest for that century. The same frequency is also significantly high compared to the long-term statistics; however such a burst of El Niño years is not entirely unprecedented. Similar levels of activity were experienced during the last three centuries of the first millennium.

This information about the long-term variability in El Niño years was then used to test the statistical significance of the trend in the frequency of El Niño years. The rate of El Niño years during the last two decades of the 20th century is about 5 years per decade. The corresponding return period, based on Figure 3.4b, is about 450 years. According to Figure 3.4a such a high frequency of El Niño years was never sustained for more than a few decades. It remains to be seen if the current episode of rather frequent E1 Niño years will be sustained for more than a few decades. If that happens, then such an episode would present a clear sign of a significant change in the global climate.

A version of this chapter first appeared as:

Eltahir, E. and Wang, G., 1999. Nilometers, El Niño, and climate variability. Geophysical Research Letters, 26(4), pp.489-492.

Chapter 4—Use of ENSO Information in Medium- and Long-Range Forecasting of the Nile Floods

Guiling Wang
Elfatih A. B. Eltahir

The Nile, running for 6,690 km, is one of the longest rivers in the world. Three main tributaries—the Blue Nile, White Nile, and Atbara River—contribute 70%, 20%, and 10%, respectively, of total Nile discharge [Hurst 1957]. Covering most of the northeastern quarter of Africa, the Nile basin has nurtured one of the most ancient human civilizations.

The Nile's flow is the main source of water in Egypt. In the late nineteenth century, from 1860–80, four serious floods caused severe property damage and loss of life in the Nile basin [Hurst 1957]. However, in the present century, the Nile basin area has been suffering from water deficits. To protect this region against droughts, as well as possible floods, the High Aswan Dam was constructed in the 1960s. The most important function of the reservoir at Aswan is to hold back water from the flood season for use in the low flow season, and from wet years to supplement the water supply in drought years.

According to the historical record, the flood peak in the Nile often occurs in September (Figure 4.1a). This causes the months June, July, and August to be a critical time for the reservoir operation. The fact that these months are also important irrigation times makes reservoir operations more complex. To make full use of the often-limited water, and to operate the reservoir in an optimal way, a high-quality medium- to long-range forecast of the Nile flood is necessary. In this chapter, we use the term “medium range” to describe forecasts with a lead time comparable to the timescale of the hydrological response, and the term “long range” to describe forecasts with a lead time longer than the hydrological response timescale. Here, the hydrological response time is the time lag between the precipitation event over the Nile catchment and the runoff delivered to Aswan station, which is about 2–3 months. The lead time is defined relative to the time of flood peak at Aswan station, which usually occurs in September.

The El Niño–Southern Oscillation (ENSO) phenomenon plays an important role in the medium- to long-range forecasts that we present in this chapter. Studies prior to the one detailed in this chapter showed that ENSO is correlated with the interannual variability of rainfall and river flow in several regions of the world. An early study by Bliss [1925] showed that Nile flow is correlated with the pressure fluctuations at Port Darwin in Australia (that is, the Southern Oscillation). The cross-spectral analysis by Richey et al. [1989] showed a significant coherency between Amazon River discharge and ENSO. According to this study, low flow in the Amazon River tends to take place

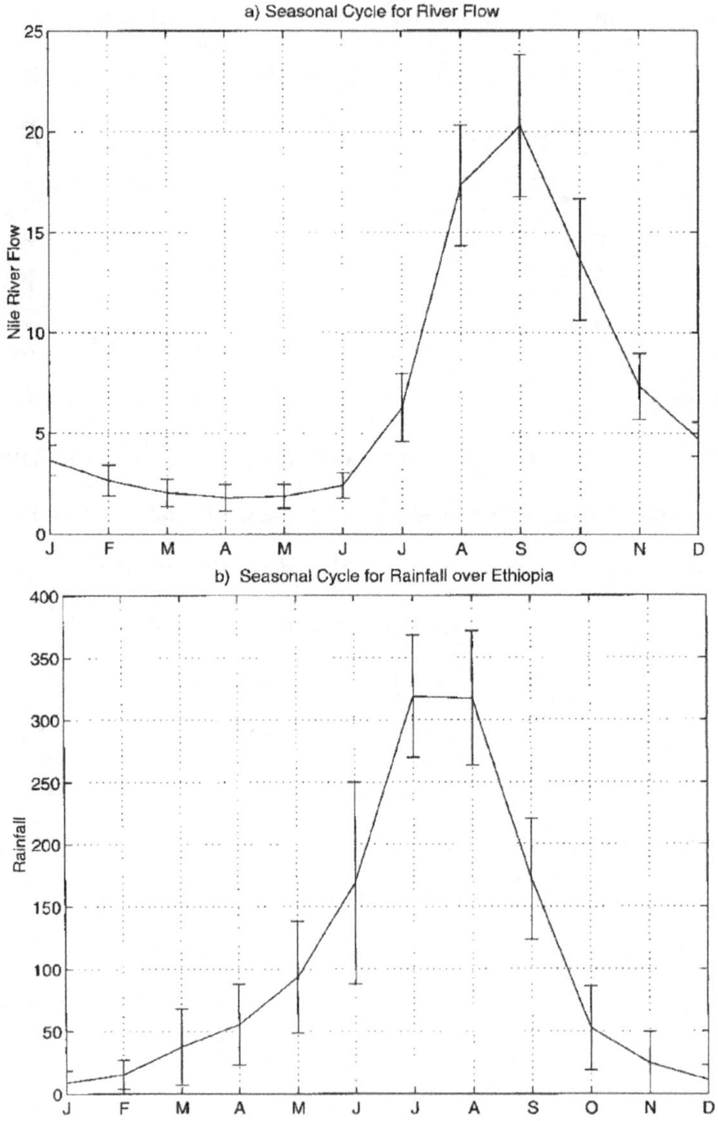

Figure 4.1a and b: The seasonal cycles for (a) Nile flow at Aswan and (b) rainfall over Ethiopia. Rainfall is measured in mm month^{-1}, and river flow is in km^3 month^{-1}.

following major El Niño events. Kahya and Dracup [1993, 1994] and Dracup and Kahya [1994] studied the influence of ENSO on the U.S. streamflow, and found that the river flow in the United States, especially in the north-central region and the Pacific Northwest, does respond to the El Niño–La Niña events. Piechota and Dracup [1996] concentrated on the droughts in the United States and found that El Niño has the strongest effect on extreme droughts in the Pacific Northwest. ENSO was also found to be correlated with rainfall and river flow in Australia by several previous studies [e.g., McBride and Nicholls 1983; Ropelewski and Halpert 1987, 1989]. The study by Ropelewski and Halpert [1987, 1989] also showed that the Ethiopia Plateau is one of the several regions in the world where interannual fluctuations of rainfall are associated with ENSO. Camberlin [1995] confirmed this result by showing that drought conditions in Ethiopia are associated with El Niño events.

The association between ENSO and the rainfall over Ethiopia contributes to the teleconnection between ENSO and Nile discharge, since the Blue Nile, which is the largest tributary of the Nile, drains the Ethiopian Plateau. According to Amarasekera et al.'s [1997] analysis, the correlation between ENSO and the discharge of the Nile is the most robust among four major tropical rivers: the Nile, Amazon, Congo, and Parana. This correlation was the basis of the proposed methodology for forecasting the Nile flood using the ENSO information.

Traditional river flow forecasting [e.g., Singh 1995] has a very short lead time, usually shorter than the hydrological response time. Since the 1980s, due to the successful ENSO predictions [Cane et al. 1986; Zebiak and Cane 1987] with a lead time of 1 year or even longer, long-range river flow forecasting has been made possible using the ENSO–river flow teleconnection. When the predicted ENSO index is used as the river flow predictor, the forecasting lead time could be much longer than the hydrological response timescale. Simpson et al. [1993a,b] used ENSO to forecast the annual discharge of Australian rivers, with a lead time up to 1 year. These studies brought a new perspective to the river flow forecasting. Similarly, as detailed in Chapter 2, the significant correlation between ENSO and Nile discharge had been examined, and ENSO used as a predictor for long-range forecasting for the Nile floods.

In all these studies, ENSO is the only predictor for the river flow. To improve the forecast skill, additional information, if available, should be incorporated into the forecasting process. Precipitation over the drainage area of the river, river flows observed prior to the forecasting time, and any other relevant factor could be valuable sources of information for the river flow forecasting. The objective of the study discussed in this chapter was to develop a forecasting algorithm that could make use of several sources of information. Our predictand here was the annual flood of the Nile and the predictors included ENSO, rainfall over Ethiopia, and the prior streamflow of the Nile at Aswan.

For any forecasting problem, there are two frequently used forecasting approaches: the linear regression and the discriminant prediction. The regression approach, with single or multiple predictors, uses the best-fit relationship to estimate the predictand, while the discriminant approach estimates the probabilities that the predictand will fall into a set of prescribed categories. The statistical techniques of these two methods are described in Parsons [1978] and Afifi and Azen [1979]. Which method to use usually depends on the particular prediction problem of interest. For purposes of operational decision-making, for example, reservoir operations, the most desirable information would be the full description of the uncertainties—that is, the probability distribution over the possible events for given values of the predictors. The operational decision can be made efficiently from a measure of the categorical probabilities [Miller 1962]. In this study, we took the discriminant prediction approach, which has been frequently used in hydrological forecasts such as the study from Chapter 2, and Folland et al. [1991] and Simpson et al. [1993a,b]. This forecasting algorithm was developed using the Bayesian theorem.

Data description

The data we used in this study include data on the Nile's flow, rainfall over the Ethiopian Plateau, and the ENSO index. The river flow data is the naturalized streamflow at Aswan, from January 1872 to December 1986. The seasonal cycle of the river flow data is shown in Figure 4.1a. The river flow in May is the lowest of the whole year, and the flow peak occurs in September. Here we define 1 June as the beginning of a hydrological year and 31 May as the end. The annual Nile flood is defined as the accumulated river flow in each hydrological year. The rainfall data is taken from a global monthly rainfall dataset, gridded at a resolution of 3.75° lat x 2.5° long [Hulme 1995]. This dataset covers the years from 1900 to 1994. We use the average in the region 7.5°–12.5°N, 37.5°–41.25°E over the Ethiopian Plateau, which covers most of the Blue Nile catchment. Figure 4.1b shows the rainfall seasonal cycle, with the maximum rainfall taking place in July. Comparison between Figs. 1a and 1b suggests that the hydrological response timescale is approximately 2–3 months. The ENSO index is derived from a monthly series of the sea surface temperature (SST) anomaly with respect to the SST climatology [Wright 1989]. Following the work describe in Chapter 2, we use the SST anomaly averaged over regions 6°–2°N, 170°– 90°W; 2°N–6°S, 180°–90°W; and 6°–10°S, 150°–110°W as the monthly ENSO index. In forecasting the Nile floods, the ENSO index averaged from September through November is used because the correlation between the Nile annual discharge and the ENSO index during September–November is the highest compared with other seasons (see Chapter 2, Table 2.1). The temporal coverage of the ENSO data is 1872–1986.

In the forecasting study, the ENSO index was used as one of the Nile discharge predictors. For the purpose of forecasting, only the forecasts made before the flood peak (in September) are useful. However, the ENSO index we used is the SST anomaly from September to November for that same year and cannot be measured until after the flood peak. Therefore, the ENSO prediction was needed in order to make real-time forecasts. Later in this chapter, we will describe our use of a predicted ENSO index (provided by Dr. S. E. Zebiak of Lamont-Doherty Geological Observatory of Columbia University, for the period 1973–86) to verify the efficiency of the ENSO prediction for the purpose of Nile flood forecasting. This predicted ENSO index is Zebiak and Cane's [1987] Niño-3 index with a lead time of 12 months.

All the data used in the forecasting study are for the period 1900–85, which is the overlapping period for all the variables.

Spectral analysis

The use of rainfall and prior river flow in the annual flood forecasting has an explicit hydrological motivation. However, the use of ENSO in the Nile flood forecasting is based on the observed teleconnection between these two variables [Ropelewski and Halpert 1987; Camberlin 1995; also discussed in Chapter 2]. Figure 4.2 plots the normalized anomalies of the Nile annual discharge and that of the September–November ENSO index. For the convenience of comparison, these two time series are plotted in

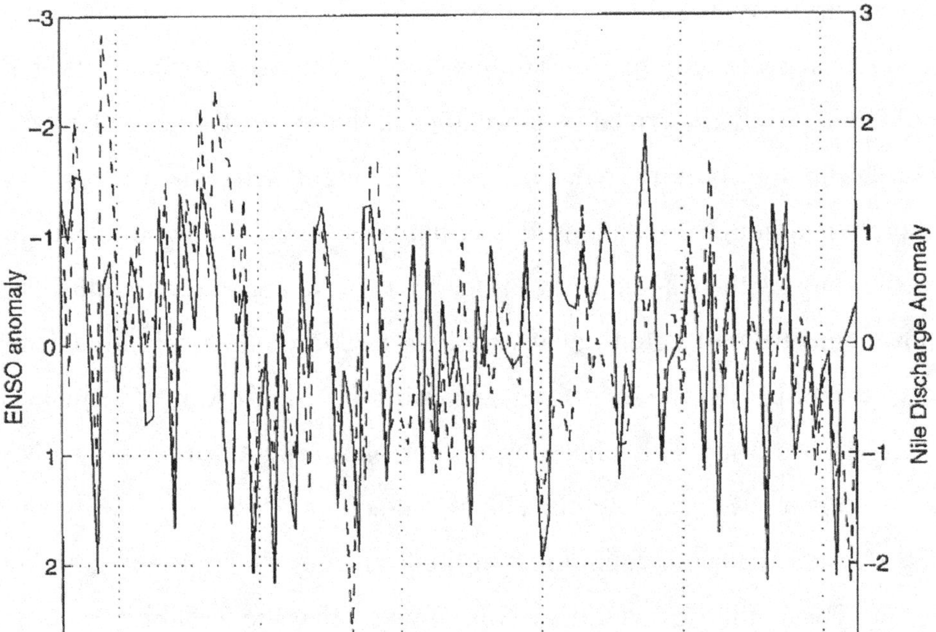

Figure 4.2: Time series of the normalized anomalies of Sep–Nov ENSO index (solid line) and the annual Nile discharge (dash line). For the convenience of comparison, the two variables are plotted using opposite coordinates. The "normalized anomalies" stand for the anomalies divided by the standard deviation.

opposite coordinates. As shown in Figure 4.2, in most of the major El Niño events, the Nile discharge is extremely low. The converse effect, high discharge associated with La Niña events, is also very obvious. The linear regression between the September–November ENSO index and the Nile annual discharge yields a correlation coefficient of -0.52. To confirm this correlation, we conducted a spectral and cross-spectral analysis. Data used in this section are the monthly Nile flow and Wright's [1989] monthly ENSO index in their overlapping period 1872–1986.

The spectrum of Nile flow and that of ENSO are shown in Figs. 3a and 3b, respectively. Consistent with previous studies [e.g., Trenberth and Shea 1987], the power spectrum of ENSO is characterized by a broad peak that extends over 3–10-year (40–120 month) periods. On the contrary, Nile flow has a distinguished peak at a period of about 1 year (12 months), although the data used here has been deseasonalized by subtracting the means and dividing by the standard deviations for different calendar months. Besides that, the power spectrum of Nile flow at very long periods is also very high. Therefore, the seasonal variability and the long-term variability in Nile flow are dominant. Figures 4.3c and 4.3d show the squared coherency and the phase spectrum between ENSO and Nile flow. As shown in Figure 4.3c, the coherence-square between Nile flow and ENSO exceeds the 1% significance level in most of the frequency domain, with a broad peak in the 3–6-yr (40–70-month) period range. This suggests that the variability of Nile flow, on the timescales of 3–6 years, is significantly correlated with ENSO. The phase spectrum (Figure 4.3d) at the corresponding timescale indicates that the anomaly of Nile flow lags

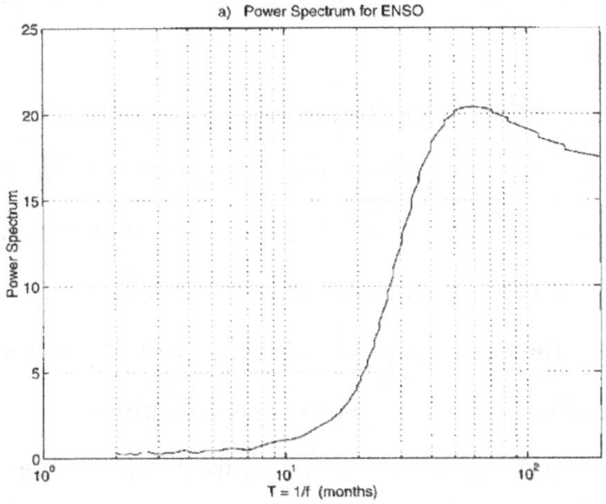

Figure 4.3a: The power spectrum of ENSO; 4.3b: the power spectrum of the Nile flow; 4.3c: the squared coherency; 4.3d: the phase spectrum between ENSO index and Nile flow. The monthly ENSO index and river flow data in the period 1872–1985 are used in these analyses. The two dashed lines in (c) show the 5% and 1% significance level, respectively.

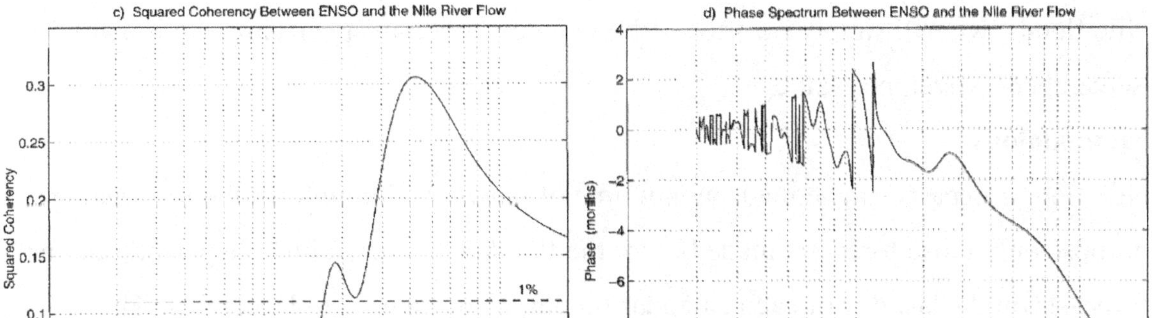

the anomaly of the ENSO index by about 3–5 months. Therefore, the ENSO index carries valuable forecasting information about Nile flow.

Although the correlation between ENSO and the worldwide river flow has long been recognized, the physical mechanism involved remains a subject for active research. Our spectral analysis presents evidence of the association between the ENSO phenomena and Nile discharge. It does not shed light on the physical mechanism behind the teleconnection. According to the cross-spectral analysis, ENSO leads Nile flow by about 3–5 months. However, as we saw in Chapter 2, the correlation between the annual discharge and the seasonal ENSO index gets its maximum in the September–November season following the Nile flood peak. This might be due to the fact that the ENSO signal is the strongest, therefore most detectable, in that season. As mentioned earlier this chapter, the averaged ENSO index from September through November will be used as a predictor of the annual Nile discharge. Hereafter, when we talk about predictors of the Nile floods, we will simply mention “ENSO index” without specifying that it is the September– November value.

Methodology

Nile flood forecasts can be made at any time of the year, although for the forecasting purpose only those forecasts made before the flood peak are useful. We performed our forecasts on the last day of each calendar month, from January to December. The forecasts after September are presented for comparison. For any forecast made before the

		Different time		
Variable 1	**Variable 2**	**Jun**	**Jul**	**Aug**
Flood	Rainfall	0.13	0.22	0.30
Flood	River flow	0.07	0.22	0.56
Flood	ENSO	0.27	0.27	0.27
ENSO	Rainfall	0.05	0.04	0.14
ENSO	River flow	0.00	0.09	0.12
Rainfall	River flow	0.00	0.05	0.21

Table 4.1: Coefficients of determination (R2) between different variables. The variable "flood" is the annual Nile flood. The variable "ENSO" is the monthly ENSO index averaged from Sep through Nov. "Rainfall" stands for the accumulated rainfall amount from May to the different forecasting time (Jun, Jul, or Aug). "River flow" is the accumulated discharge from Jun to the forecasting time.

beginning of the hydrological year, no rainfall or river flow information about that hydrological year is available, so we simply used the latest monthly rainfall and river flow observations available as the sources of information. For the forecasts made after 1 June, the total river flow from the beginning of June accumulated up to that time is used as one predictor. Due to the time lag between rainfall and river flow, rainfall in May of the last hydrological year also contributes to the annual flood of the current hydrological year, so the rainfall information used is the total rainfall accumulated since the beginning of May.

The correlations of annual Nile flood with corresponding accumulated rainfall and river flow at different times are analyzed and tabulated in Table 4.1. The first two rows of Table 4.1 list the coefficients of determination between the annual flood and the amount of rainfall and river flow used in the forecasts on the last day of June, July, and August. As the time moves further into the hydrological year, the correlation coefficients become larger, therefore the rainfall and river flow information become more useful for the forecasting process. When the forecasting time moves toward the end of the hydrological year, it is obvious that the coefficient of determination between the accumulated river flow and the annual flood will approach 1.0.

The discriminant prediction we used here is also called "categoric prediction," which forecasts the categoric probabilities of the predictand according to the categories that the predictors fall into. All the data points are divided into three categories, as is done in many of the previous studies [Simpson et al. 1993a,b; and Chapter 2]. The data of Nile flow and annual flood are classified into "low," "normal," and "high" categories. The rainfall data categories are labeled "dry," "normal," and "wet." Similarly, the ENSO index is categorized as "cold," "normal," and "warm." For the Nile flood, rainfall, and prior river flow, the definition of normal is the range from $\mu - \sigma/2$ to $\mu - \sigma/2$ (μ is the mean value and σ is the standard deviation).

The other two categories contain those data lower and higher than normal. For the ENSO index, we follow Simpson et al. [1993b] and the information presented in Chapter 2 to use −0.5°C and 0.5°C as the boundaries for the "normal" category. Any condition with an ENSO index lower than −0.5°C is defined as cold, and higher than 0.5°C as warm. The way we define the boundaries between different categories is not totally arbitrary. It is designed in such a way that the number of data points contained in every category of a specific variable are not dramatically different from each other. In the real forecasts, the division of the categories, even the number of the categories, depends on the particular forecast purpose. For example, when only the extreme events (e.g., extremely wet or dry events, or the extreme El Niño or La Niña episodes) are considered, a much wider range for the "normal" category or a larger number of categories are recommended.

The categoric probabilities for the predictand—in this case the annual flood in the Nile river—will be the forecast results. For the forecasts that have three predictors, the categoric probability is the conditional probability with three conditions; that is, the probability of having a Nile flood in any category (low, normal, and high) given a certain category of ENSO condition (cold, normal, or warm), a certain category of rainfall condition (dry, normal, or wet), and a certain category of river flow condition (low, normal, or high). (To compute these conditional probabilities, we developed an algorithm using the Bayesian theorem, which will be described later in this chapter). For a specific year, the forecasting results would be the three probabilities for the annual flood falling into each of its three categories.

However, generally, the forecast is expressed by the forecasting probabilities of each category of the Nile flood given every possible combination of ENSO, rainfall, and river flow conditions. To easily judge the forecasts skill and to compare different forecasts, we proposed a synoptic parameter, the forecasting index (FI), which can measure the forecast skill not only in this study but also in any other discriminant forecast. The FI value is defined as the average of the forecasting categoric probabilities for the categories that the observed Nile flood in each individual year falls into during an n-yr period. First of all, in each year j ($j = 1, \ldots, n$), the ENSO index, rainfall, and river flow data are categorized. Accordingly, the probability of each Nile flood category can be forecasted, and we denote it as the prior probability $P_r\,(i, j)$, $i = 1, 2, 3$. Then, the Nile flood observation in that year is categorized, and the posterior probability $P_p\,(i, j)$ can be identified as [1, 0, 0] in a low flow year, [0, 1, 0] in a normal year, and [0, 0, 1] in a high flow year. FP(*j*), the forecasting probability of the Nile flood category, which describes the observed flood condition for that year, can be computed as

$$\mathrm{FP}(j) = \sum_{i=1}^{3} P_r(i,j)P_p(i,j)$$

The forecasting index, FI, is the average of these probabilities over a certain period:

$$\mathrm{FI} = \frac{1}{n}\sum_{j=1}^{n} \mathrm{FP}(j) = \frac{1}{n}\sum_{j=1}^{n}\sum_{i=1}^{3} P_r(i,j)P_p(i,j) \qquad (4.1)$$

A larger forecasting index implies a more accurate forecast. A perfect forecasting methodology would have an FI of 1.0.

Traditionally, the discriminant forecast skill is measured using the proportion of correct forecasts, in which the right category of the predictand gets the highest predicted probability. Therefore, when the traditional skill measure is used, a forecast of 80% (wet), 15% (normal), and 5% (dry) for a wet year is no difference from a forecast of 40% (wet), 35% (normal), and 25% (dry). However, the skill measure proposed in this study can tell the “accuracy” difference between different forecasts, even if they might be all correct forecasts. This new measure of skill provided the field with a better tool to judge the quality of categoric predictions.

The Bayesian theorem and forecasting algorithm

Algorithm development

Any conditional probability can be computed by counting all the relevant data points and normalizing it by the number of all data points that satisfy the condition or conditions. For this study, we were interested in the conditional categoric probabilities. All the data points that fulfill the given condition(s) were counted; we then identified those cases with a low, normal, and high annual flood, and got the relative frequency distribution. Then the forecasting probability of each flood category could be computed. By counting the data points conditionally, the total data points were divided into several groups. If only one condition was considered, the relevant data points were divided into nine groups, which is shown by Figs. 4a–c. To avoid statistical error, the number of the total data points should be much larger than the number of groups into which the whole record is divided by conditional counting. However, the number of groups increases exponentially with the number of conditions. Considering only two conditions results in 27 groups, and with three conditions we get 81 groups. Usually, the limited data size makes this process impractical.

In this study, we dealt with the forecasting probability of the Nile flood using three sources of information (ENSO, rainfall, and prior river flow). Therefore, a conditional probability with three conditions was involved. To break through the limitation of data size, we developed an approximate algorithm using the Bayesian theorem. Details for Bayesian theory are described in many statistical books (e.g., Winkler 1972; West 1989).

The Bayesian theorem can be expressed as

$$P(Q_i/A,B) \;=\; \frac{P\left(\frac{B}{Q_i},A\right)P\left(\frac{Q_i}{A}\right)}{P\left(\frac{B}{A}\right)} \qquad (4.2)$$

where $P(Q_i/A)$ is the possibility of event Q_i given that event A has occurred, and $P(Q_i/A,B)$is the possibility of event Q_i given that both event A and event B have occurred. Here, $P(B/Q_i,A)$, $P(B/A)$ and the following probabilities can be defined in a similar manner. Assuming A and B are independent events, we can rewrite Equation (4.2) as

$$P(Q_i/A,B) = \frac{P(B/Q_i)P(Q_i/A)}{\sum_{j=1}^{3} P(B/Q_j)P(Q_j/A)} \qquad (4.3)$$

where Q_j represents a full group of events. Equation (4.3) expresses $P(Q_i/A,B)$ as a function of $P(Q_i/A)$ and $P(B/Q_i)$, $i = 1, 2, 3$. When we compute $P(Q_i/A)$, $P(B/Q_i)$by the counting procedure, we only need to divide the available data into nine groups. This can significantly reduce the statistical error resulting from a limited data size. The advantage of assuming independence in the Bayesian algorithm is that the records of the variables defining events A and B can be totally nonoverlapping. If we have the data on Q_i for both period 1 and period 2, and have data on A only in period 1 and data on B only in period 2, the conditional probability can still be computed using Equation (4.3), because no relation between A and B is needed. This allows us to include more information beyond the overlapping period when the data in the overlapping period is too limited.

For the case including a third independent condition C, the conditional probability $P(Q_i/A,B,C)$can be derived from Equation (4.3).

Assume $D = A \cup B$; then C and D are independent since A, B, and C are independent. Therefore,

$$P(Q_i/D,C) = \frac{P(C/Q_i)P(Q_i/D)}{\sum_{j=1}^{3} P(C/Q_j)P(Q_j/D)}$$

Substituting $D = A \cup B$ into the above equation yields

$$P(Q_i/A,B,C) = \frac{P(C/Q_i)P(Q_i/A,B)}{\sum_{j=1}^{3} P(C/Q_j)P(Q_j/A,B)} \qquad (4.4)$$

The conditional probability with more conditions can be derived similarly.

By substituting Equation (4.3) into Equation (4.4), a three-variable conditional probability can be computed from a set of one-variable conditional probabilities. If Q, E, P, and R stand for the Nile annual flood (Q), the ENSO ***(E)*** index, precipitation over Ethiopia (P), and prior Nile River flow (R), respectively, the forecasting probability $P(Q_i/E,P,R)$ is a function of $P(R/Q_j)$, $P(P/Q_j)$, $P(Q_j/E)$, $j = 1, 2, 3$.

To use Equations (4.3) and (4.4) for this study, we needed to assume the independence between the input variables E, P, R. Although these three variables are correlated with one another, before September, the mutual correlation between them is relatively weaker than their correlations with the Nile annual flood, as shown in Table 4.1, which uses June, July, and August as examples. The upper half of the table lists the coefficients of determination between the annual Nile flood and each of the input variables; the lower half lists the coefficients of determination between each of the three

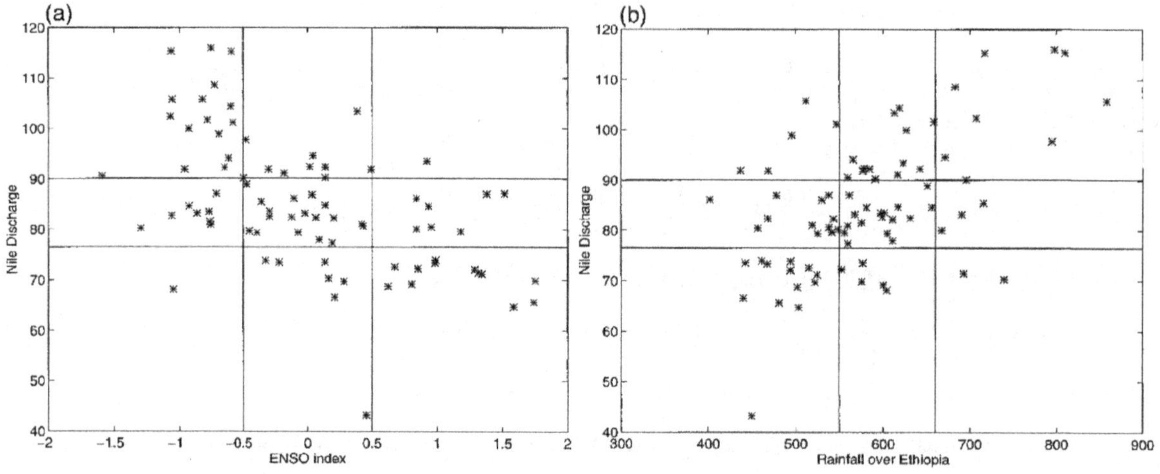

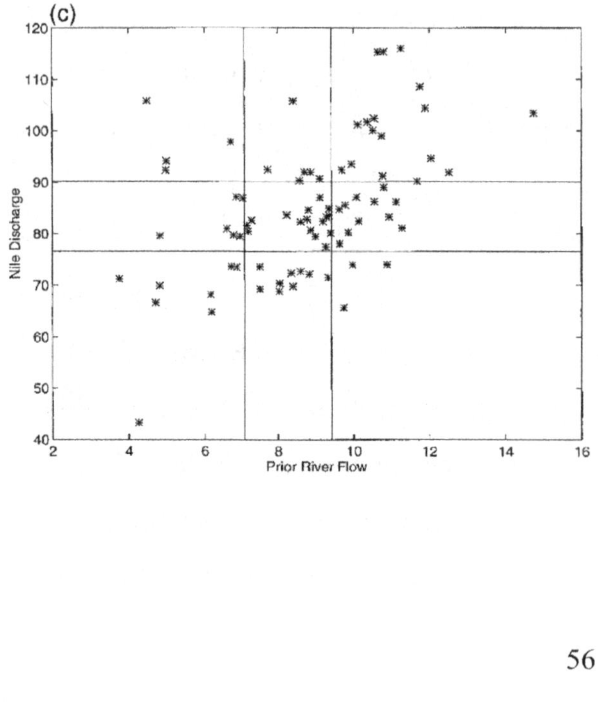

Figure 4.4: Counting the data on the Nile flood with one condition (a) ENSO, (b) rainfall, (c) river flow, respectively, divides the available data into nine minor groups. ENSO is measured in 8C, rainfall in mm; and river flow in km3. Rainfall in (b) is the accumulated amount from May to July. The prior river flow in (c) is the accumulated discharge from June to July.

Conditions		Flood		
ENSO	**Rainfall**	**Low**	**Normal**	**High**
Cold	Normal	0.14/0.02	0.64/0.66	0.21/0.32
	Wet	0.00/0.01	0.12/0.00	0.88/0.99
Normal	Dry	0.58/0.49	0.42/0.44	0.00/0.07
	Normal	0.33/0.19	0.61/0.73	0.06/0.08
Warm	Dry	0.77/0.79	0.23/0.19	0.00/0.02
	Normal	0.50/0.48	0.50/0.48	0.00/0.04

Table 4.2: Nile flood forecast based on ENSO and rainfall information. The probabilities over the slash are derived using data counting, and those under the slash are the forecast results using the Bayesian algorithm.

input variables. In every month, the values in the lower half are much smaller than those in the upper half. Thus we can assume that the input variables are statistically independent, although physically they are not.

Effect of the independence assumption

To look at the influence of the assumption of independence between the predictors, a simple experiment was designed. The data used covers the period 1900–85. If only two predictors were considered, this record length is adequate for using the data-counting method to compute the conditional probability of the Nile flood. We forecasted the Nile annual flood using any two of these three predictors (the ENSO index, rainfall, river flow) by applying both the Bayesian algorithm [Equation (4.3)] and the data-counting procedure. As an example, Table 4.2 tabulates the results of the forecasts based on ENSO and rainfall information conducted in July. We listed only the forecasting probabilities in the categories within which we have enough data points for the use of the data-counting procedure. For the other combinations of rainfall and ENSO categories, we had too few data points (less than 5 points) to conduct the counting procedure. In Table 4.2, the value

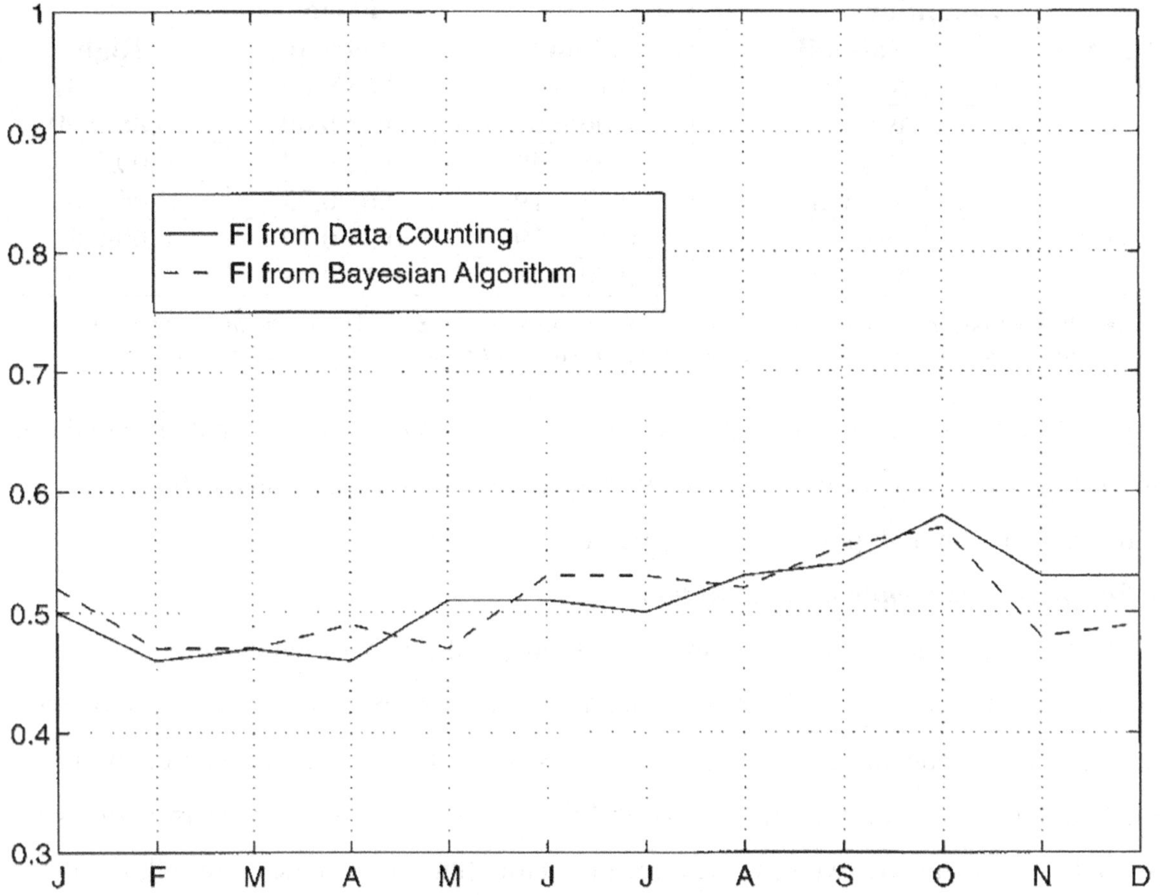

Figure 4.5: The FI values for forecasts based on ENSO and rainfall information using two different methods: data counting and the Bayesian algorithm.

over the "/" is derived in the data-counting procedure, and that under the "/" is derived using the Bayesian algorithm. As shown, there is very small difference in terms of the relative magnitude of the three probabilities; that is, the category that gets the highest probability in the datacounting process still has the highest probability when it comes to the Bayesian algorithm. The forecast skills measured by the FI values for the two methods are plotted in Figure 4.5, which shows no systematic difference between these two.

Figure 4.5 might imply that for the dataset size and correlation magnitude we dealt with in this study, the statistical errors due to the data size limitation for the two-condition data-counting method had a similar magnitude as the errors caused by the independence assumption in the algorithm we developed in this study. In other words, the advantage brought to the forecasting process by breaking through the limitation of data size could compensate for the disadvantage caused by the approximate assumption. Comparison between the two methods for forecasts using other combinations of predictors (i.e., rainfall and river flow, or ENSO and river flow) supports the same conclusion. The benefit from Bayesian methodology is larger when we deal with three predictors. Therefore, in this study, the damage for the forecasts quality caused by the mutual dependence between ENSO index, rainfall over Ethiopia, and Nile flow could be small. Hence, we used Equations (4.3) and (4.4) in our study without considering the mutual dependence between the three predictors.

Results

Using the Bayesian algorithm, we forecasted the Nile annual flood using information about ENSO, rainfall, and river flow as predictors. We conducted both a calibration and a verification analysis. The calibration period covers 50 years, from 1900 to 1949, and the verification period is 36 years-long, from 1950 to 1985.

Predictors			**Predictand (annual flood)**		
ENSO	**Rainfall**	**River flow**	**Low**	**Normal**	**High**
Cold	Dry	Low	0.33	0.44	0.23
		Normal	0.25	0.65	0.10
		High	0.10	0.48	0.42
	Normal	Low	0.05	0.45	0.50
		Normal	0.04	0.72	0.24
		High	0.01	0.34	0.65
	Wet	Low	0.04	0.16	0.80
		Normal	0.05	0.37	0.58
		High	0.01	0.10	0.89
Normal	Dry	Low	0.66	0.29	0.05
		Normal	0.53	0.45	0.02
		High	0.34	0.52	0.14
	Normal	Low	0.21	0.59	0.20
		Normal	0.14	0.80	0.06
		High	0.06	0.59	0.35
	Wet	Low	0.47	0.29	0.24
		Normal	0.23	0.52	0.25
		High	0.06	0.25	0.69
Warm	Dry	Low	0.89	0.11	0.00
		Normal	0.82	0.18	0.00
		High	0.72	0.28	0.00
	Normal	Low	0.59	0.41	0.00
		Normal	0.42	0.58	0.00
		High	0.29	0.71	0.00
	Wet	Low	0.77	0.23	0.00
		Normal	0.63	0.37	0.00
		High	0.49	0.51	0.00

Table 4.3: The Nile flood probabilities for July forecast using ENSO, rainfall, and river flow information.

Our main objective here was to forecast the Nile flood using three sources of information, the ENSO index, rainfall over Ethiopia, and prior Nile flow. To show the improvements that these three sources of information bring to the quality of the flood forecasting, the corresponding forecasts using two predictors, one predictor, and no predictor were also performed. The case with no predictor refers to a forecast based solely on the historical record of the annual flood, which is determined by the division of the annual discharge categories. In this case, using 1900–49 as the calibration period, the forecasting probabilities for the low, normal, and high flow categories are 30%, 44%, and 26%, respectively. The forecasts using the predictor(s) start by adding ENSO information, then adding the rainfall and river flow information. To show the importance of ENSO information, we also performed the forecasts using only rainfall and river flow as predictors. Although only the forecasts made before the Nile flood peak are useful, for comparison we also performed those after the time of the flood peak. For every case, the forecasting time extends from the end of January to the end of December. All the forecasts are made at the end of every month. To be compact, hereafter, we will only mention the name of that month when referring to the forecasts made at the end of the month.

Here we use July as an example to show the calibrated forecasting probabilities with three predictors, as tabulated in Table 4.3. Once the ENSO, rainfall, and prior river flow information for the July forecasts are collected, Table 4.3 will give the forecast probabilities. Suppose that in a specific year, the ENSO index is categorized as cold, and rainfall and river flow as wet and high, then the probabilities for the annual discharge to

Predictors			Forecast type	Predictand (annual flood)		
ENSO	Rainfall	River flow		Low	Normal	High
Cold	Wet	High	1	0.30	0.44	0.26
			2	0.07	0.40	0.53
			3	0.01	0.10	0.89
Warm	Dry	Low	1	0.30	0.44	0.26
			2	0.67	0.33	0.00
			3	0.89	0.11	0.00

Figure 6: Probabilities of the Nile flood in several extreme cases of the Jul forecasts. Type 1 stands for the forecasts based on the historical record; type 2 stands for the forecasts using ENSO as the predictor; type 3 stands for the forecasts using ENSO, rainfall, and river flow as predictors.

fall into the low, normal, and high categories are 1%, 10%, and 89%, respectively. In Table 4.3, it is worth noticing that the probability of having a high Nile flood is always zero when the ENSO index is in warm category, which means that no high flood has ever been observed during major El Niño events in our calibration period. In Table 4.4, the forecasting probability for two extreme cases from Table 4.3 are compared with the forecasting results obtained by using the historical record and by using ENSO solely. It shows that the addition of ENSO information to the information from the historical record modifies the forecasting probability for the Nile flood to a large extent. Significant modification is also obtained by introducing the rainfall and prior river flow information in addition to ENSO. In both cases, the incorporation of additional information tends to result in a more definite forecast; that is, one of the three categoric probabilities stands out and the other two shrink.

Figures 4.6a and 4.6b demonstrate the skill measurements (FI values) of the different forecasts. As we explained earlier in this chapter, the FI value is computed by comparing the forecasts with the observations.

Calibration (1900–49)

Figure 4.6a shows the FI values for different forecasts in the calibration period. Based on the historical flood record, we obtain an FI value of about 0.35. Because the Nile flood data are divided into three categories, any FI value not significantly higher than one-third indicates a trivial forecast. The ENSO information introduces a significant improvement, with a jump of FI value as high as 0.1, from 0.35 to 0.45. It is worth noting that the skill improvement resulting from ENSO information does not change with the forecasting time: FI is as high as 0.45 (solid line with "+" sign) even as early as January. If the rainfall and river flow information are used in addition to the ENSO information, we get a forecasting index that changes with the forecasting time. Before June, which is the beginning of the hydrological year, the addition of rainfall and river flow information improves the forecasts skill by about 0.05. Skill comparison between the forecasts using all three predictors (solid line) and the forecasts using ENSO and rainfall (dash–dot line) shows that both the rainfall and river flow information contribute to this improvement. The dashed line in Figure 4.6a represents the skill for the forecasts based on rainfall and river flow information. Before May, using rainfall or river flow as predictors is equivalent to assuming pure hydrological persistence. As shown, the forecast skills based on persistence (dash line) are much lower than those based on ENSO index (solid line with "+" sign).

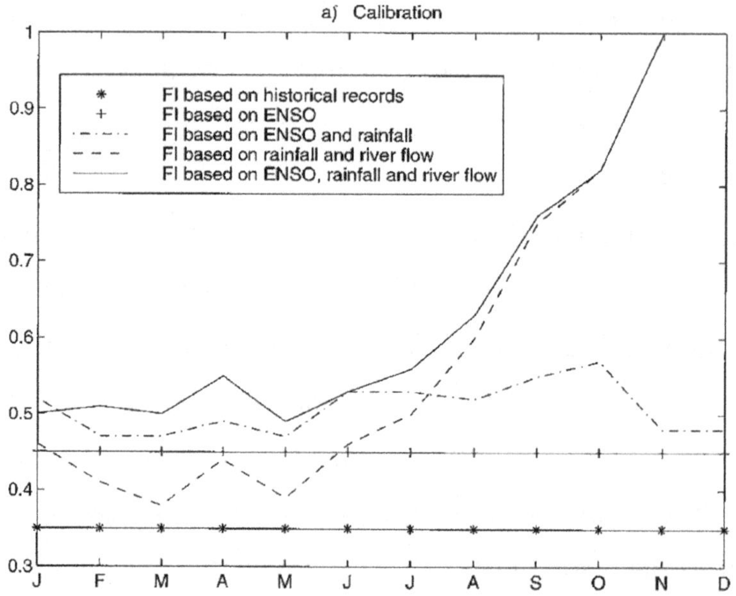

Figure 4.6: The FI values in (a) calibration period (1900–49) and (b) verification period (1950–85).

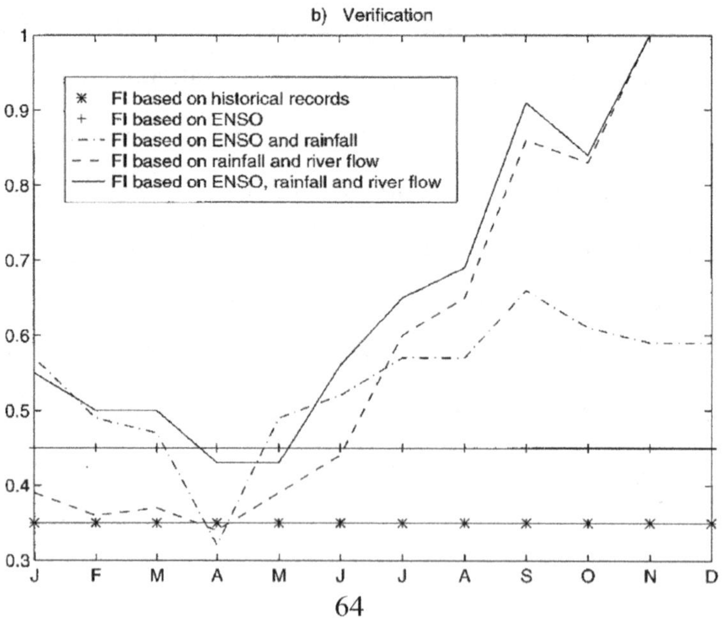

All these imply that for long-range forecasting, ENSO is the most important source of information. The rainfall and river flow information available before June are from the last hydrological year, so their role in the long-range forecasting of the flood in the current hydrological year is not as important as ENSO.

Since June, with more information about rainfall and river flow becoming available, the forecasting quality gains a larger improvement from the addition of rainfall and river flow information. The rainfall information boosts the FI value from 0.45 (solid line with "+" sign) to 0.53 (dash–dot line), which does not change significantly in the forecasts made from June to September. River flow information introduces additional improvement to the forecasting quality, which becomes higher and higher as the forecasting time progresses into the current hydrological year. These lead to an FI value for the forecasts using three predictors (solid line) as high as 0.56 in July and 0.63 in August. Notice that the Nile flow peak occurs in September. So, on average, 2 months before the flood peak, we forecast a probability of 56% for the correct flood category, and 1 month before the flood peak, this forecasted probability is 63%. Comparing with the FI value for forecasts using ENSO information solely (0.45), we concluded that the rainfall and river flow information are very important in improving the forecasting quality of the medium-range forecasts. ENSO is also an important factor in the medium-range forecasting since there is still a big gap between the skills for forecasts using rainfall and river flow information and the skills for forecasts using ENSO, rainfall, and river flow information.

After the time of the flood peak, significant information about the river flow is available, thus it becomes almost trivial to identify the right category of the annual flood. As expected, the forecasting index approaches 1.0 when the river flow information is added to the forecasts.

Verification (1950–85)

The forecast skills in the verification period are shown in Figure 4.6b. Similar to the results in the calibration period, ENSO is the most important information source for the long-range forecasts, and rainfall and river flow information improve the medium-range forecast skill. Two noticeable differences between the calibration and verification periods are observed:

1) In the verification period, there is a dip in the skill of April and May forecasts. This is caused by some change in the rainfall seasonal pattern over the Ethiopian Plateau. During the verification period, there are several extremely dry springs, with rainfall amount less than 20% of its normal value, followed by a very wet or normal summer. This kind of rainfall pattern, or the other way around (wet spring followed by dry summer), was not observed in the calibration period. This change of precipitation persistence causes the low forecast skill in April and May of the verification period.

2) The medium-range forecast skills in the verification period are generally higher than those in the calibration period. The FI value for forecasts using three predictors is 0.65 in July and 0.69 in August, compared with 0.56 and 0.63 in the calibration period. The FI value for the forecasts using ENSO solely is the same as in the calibration period.

Therefore, rainfall and river flow bring more information to the medium-range forecasting in the verification period than in the calibration period.

As cited before, an FI not larger than 0.34 indicates a trivial forecast because the Nile flood is divided into three categories. It is worth noting that, because of the same reason, an FI value equal to or larger than 0.5 indicates a successful forecast. For example, suppose that in a wet year, the forecast probabilities for the three Nile flood categories are A (wet), B (normal), and C (dry), where $A + B + C = 1.0$. If $A > 0.5$, both B and C must be less than A. Therefore, a correct forecast is guaranteed regardless the values of B and C. For our forecasts with three predictors, the FI values were almost always larger than 0.5 in both the calibration and the verification periods.

Effect of ENSO prediction

In the above "forecasts," the ENSO index is based on observation. However, in practice, we need to use the predicted ENSO index [SST anomaly in the tropical eastern Pacific Ocean (TEP) from September to November]. To investigate the influence of ENSO prediction on Nile flood forecasting, we compared the forecasts based on the ENSO prediction ("Pre." forecasts) with the hindcasts based on the ENSO observation ("Obs." forecasts). The predicted ENSO index covers the period 1973–85. Figure 4.7 plots the FI values for the forecasts with three predictors, using 1973–85 as the verification period. The dashed line stands for the Obs. forecasts, and the solid line stands for the Pre. forecasts. Similar to Figure 4.6b, the low forecast skills in April and May are caused by the shift of the rainfall seasonal pattern. Figure 4.7 shows no significant difference in the FI value for the forecasts using ENSO prediction and ENSO observation, which suggests

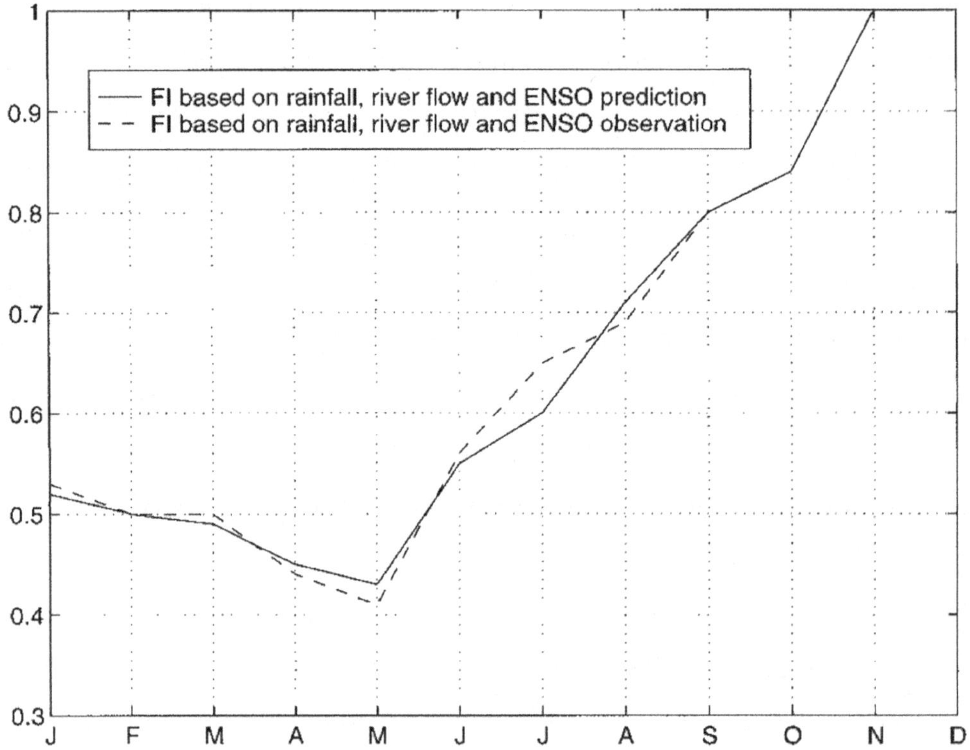

Figure 4.7: FI value in a verification period 1973–86 for forecasts with three predictors using predicted ENSO index (solid line) and observed ENSO index (dashed line). The predicted ENSO index is Zebiak and Cane's (1987) Niño-3 index; the observed ENSO index was derived from Wright's SST data.

that the predicted ENSO index works as well as the observed one. The long-range forecasting of the Nile flood can rely on the predicted ENSO information.

Conclusions

The correlation between ENSO and the Nile discharge was the basis of this proposed methodology for the medium- to long-range forecasting of the Nile flood. The ENSO index solely yields a reasonably accurate forecast for the Nile flood, with an FI value of 0.45, compared to 0.35 for the forecast based solely on the historical record. In this study, we developed an algorithm to combine other information with ENSO, and performed medium- to long-range forecasting of the Nile flood using information about ENSO, rainfall over Ethiopia, and prior river flow of the Nile. For the long-range forecasting of the Nile flood (lead time longer than the hydrological response timescale), ENSO information plays the dominant role. However, for the medium-range forecasts (lead time shorter than the hydrological response timescale), information about rainfall and prior river flow brings large improvement to the forecasting quality. In June, July, and August, the FI values for the forecasts using three predictors are about 0.10–0.20 higher than the FI value for the forecasts using ENSO solely. By using these additional information sources, we achieved a significantly better medium-range forecast.

As discussed previously, an FI value larger than 0.5 implies a successful forecast since the forecast probability for the correct category is the highest of the three. The skill increase caused by the rainfall and river flow information, ranging from 0.05 to 0.20 in this study, boosted the FI value over 0.50 in almost all the forecasts using three predictors. Therefore, we view the improvement due to rainfall and river flow information as significant. Our medium- to long-range forecasts for the Nile river annual flood were successful.

For practical purposes, the predicted SST in TEP should be used as the ENSO index. Based on the results of this study, the ENSO prediction is good enough to be used in the medium- to long-range forecasting of the Nile flood. This assures the practicality of the forecasting procedure proposed in this chapter.

A version of this chapter first appeared as:

Wang, G. and Eltahir, E., 1999. Use of ENSO Information in Medium- and Long-Range Forecasting of the Nile Floods. Journal of Climate, 12(6), pp.1726-1737.

Chapter 5—Climate Change Enhances Interannual Variability of the Nile River Flow

Mohamed S. Siam
Elfatih A. B. Eltahir

The Nile river basin is an ecosystem under severe stress. The human population living in the Nile basin countries is projected to double by 2050, approaching one billion [United Nations, 2013]. The basin is currently shared by about 400 million people in eleven countries with economies that depend heavily on agriculture, which employs the vast majority of the labor force in most of these countries [United Nations, 2013]. Furthermore, almost half of the Nile basin countries are projected to live below water scarcity level, 1000 m3/person/year, by 2030 [Baecher, et.al., 2015] [FAO, 2000]. Thus, any future changes in the magnitude of the flow volume of the Nile River can lead to significant impacts on the lives of people living within the basin and may increase the already high level of water stress. In order to fully utilize the water resources of the basin, several dams were built in the last century to control the seasonal and inter-annual variability of the Nile flow.

NATURAL VARIABILITY OF THE NILE FLOODS

The recent conflict over the Nile water has received significant attention in the last few years after the decision by Ethiopia to build a large dam on the Blue Nile (the Grand Ethiopian Renaissance Dam, or GERD) to produce electricity, mostly for export to neighboring countries.

The dam, which began construction in April 2011, is relatively large compared to previous designs for the same location. This raised serious concerns regarding its effect on water shares of downstream countries (i.e. Egypt and Sudan). If variability of the Nile flow changes in the future, then water storage capacity in the basin will need to be revaluated.

Until recently, attempts to project the future of the Nile flow yielded inconsistent results. While several studies examined the impacts of climate change on the Nile basin using different approaches, high uncertainty surrounded conclusions from these studies for several reasons. First, none of the previous studies presented observational evidence to support their hypotheses as they estimated the impacts of climate change on the flow of Nile river only using Global Climate Models (GCMs) without presenting recent observations that show similar trends. Second, no consistent physical mechanism was advanced to explain the future changes. Third, wide disagreement persists between the results of simulations using the GCMs, not only in magnitude but also in the sign of projected changes in rainfall over the basin [Christensen 2007, 2013]. These combined factors made it difficult to reach a consensus on the impact of climate change on the Nile river flow. In this chapter, we use recent observations and climate model simulations to

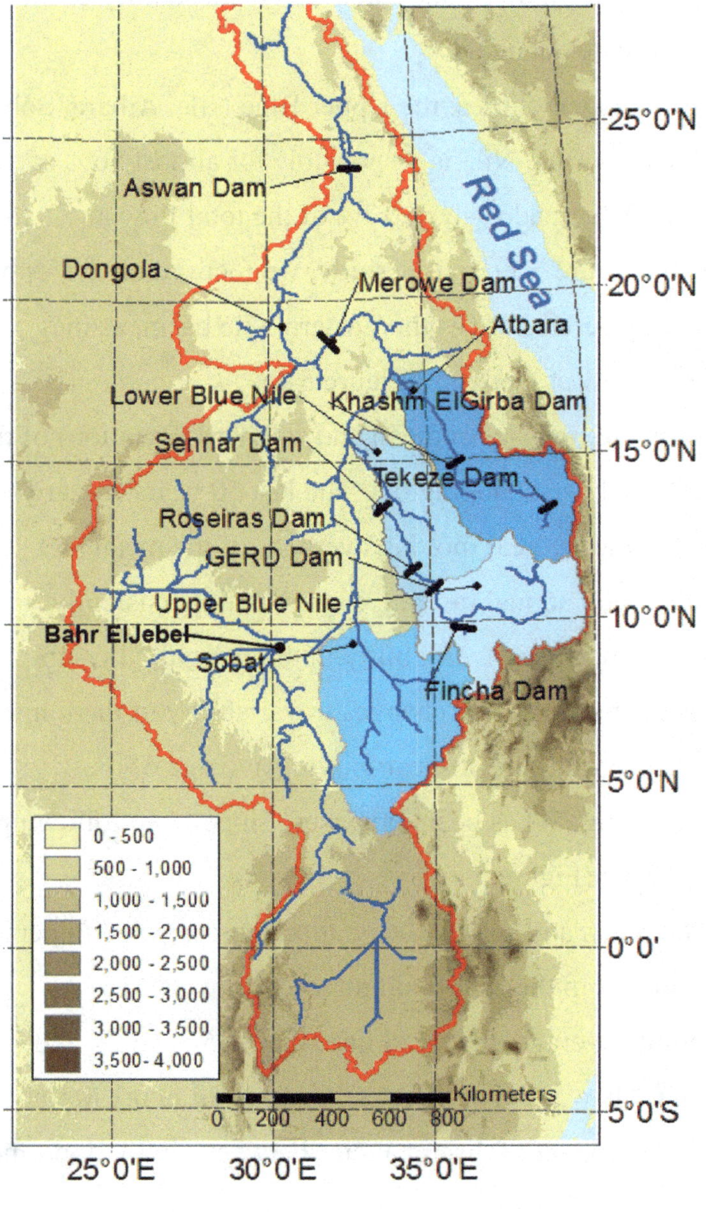

Figure 5.1: Topographic map of Eastern Africa and the Nile sub-basins. Topographic map of Eastern Africa showing the main tributaries of Nile basins (Upper Blue Nile, Sobat, Atbara and Bahr El Jebel) and different dams in these basins. The rainfall and runoff data analyzed in this paper are averaged over the Upper Blue Nile, Sobat and Atbara basins.

support the hypothesis that both the long-term mean and inter-annual variability of the Nile river flow may increase due to climate change.

The Nile river basin has four main tributaries: the Upper Blue Nile, Atbara, Sobat and Bahr el-Jebel (Figure 5.1). The Upper Blue Nile is responsible for almost 56% of the total Nile flow, and each of Sobat and Atbara adds about 12% of the total flow at Dongola [Conway and Hulme, 1993] (Figure 5.1). In this study, we focused on the Upper Blue Nile, Sobat and Atbara, hereinafter referred to as the Eastern Nile basin, as they represent almost 80% of the total Nile flow at Dongola (Figure 5.1).

The flows and rainfall patterns (i.e. long-term mean and standard deviation) of the Upper Blue Nile, Sobat and Atbara have been changing over the last 50 years (Figure 5.2a-c). These changes are reflected in the 30-year moving averages of the mean and standard deviation of annual river flow and annual rainfall, as they both persistently increase with time (Figure 5.2d-f). The increase of variability in the flow of Nile sub-basins during the last 50 years is also consistent with changes in variability of the longer record describing main Nile flow at Dongola. The 30-year moving average of the coefficient of variation of the flow at Dongola has been increasing since 1940 following a previous peak in variability around 1920s (Figure 5.3c).

The observed increases in the mean and standard deviation are projected to persist during the twenty-first century, as evident in the runoff simulated by the GCMs (indicated by the multi-model ensemble average, thick red line, Figure 5.3a, 3.3b and 3.3d). The runoff simulated by the GCMs was bias-corrected, as outlined in the Methods section below, in order to minimize the impact of discrepancies between simulations and

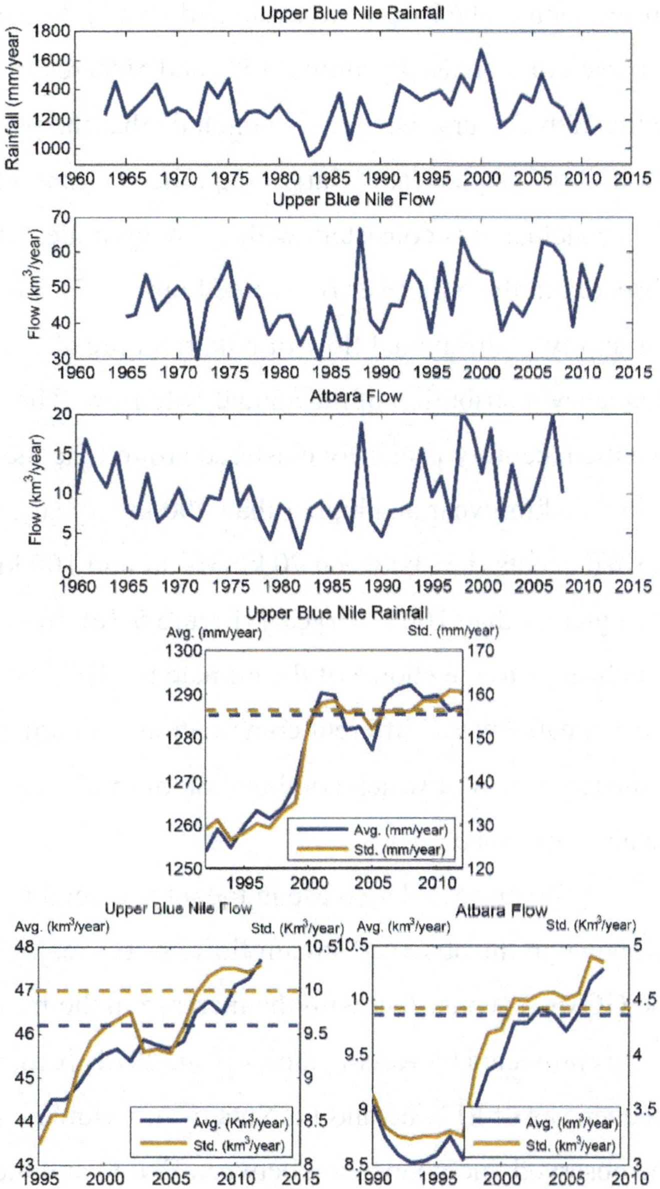

Figure 5.2: Observed stream flows and rainfall, and moving averages for the mean and standard deviation for the Upper Blue Nile and Atbara basins. a, time series of the annual rainfall based on the weighted average of rainfall stations over the Upper Blue Nile basin. b, c, time series of the annual stream flow averaged between June and May for the Upper Blue Nile and Atbara basins, respectively. d, 30-year moving averages for the rainfall mean and standard deviation of the Upper Blue Nile. e, f, 30-year moving averages for the stream flow mean and standard deviation of the Upper Blue Nile and Atbara basins respectively. In d, e, f, the dashed lines in blue and brown are the 90% confidence levels of the mean and standard deviation using the t-test and Chi-square distributions, respectively, based on the mean and standard deviation of the periods (1965-1995) and (1960-1990) for the Upper Blue Nile and Atbara, respectively. In d, e, and f the moving average is plotted at the last year of the 30 years.

observations. The long-term mean and standard deviation of total annual river flow are expected to increase by almost 15% and 50% respectively for the 21st century compared to the 20th century. Hence, we conclude that the projected relative change in inter-annual variability is significantly larger than the corresponding change in long term mean flow. This conclusion is consistent with the magnitudes of the changes that has already been observed in the Nile river flow (See Figure 5.2). These changes in the mean and variability in the annual flow of Nile river are also evident in the projected change in the frequency distribution of the annual Nile flow. The distribution of the flow in the twentieth century is closely clustered around the mean with few extreme events (i.e., below 70 km3/year and higher than 100 km3/year), while the future flows have fewer normal events (i.e., between 70 km3/year and 100 km3/year) and more high flow events (i.e., greater than 100 km3/year) (Figure 5.3e). In order to study the sensitivity of our conclusions to the choice of the climate models, the projected changes in the Nile flow are estimated using different combinations of climate models. Our conclusions remain the same regardless of which combination of models is chosen to estimate the trends in the mean or variance.

The projected increase in the inter-annual variability of the flow of the Nile river, as shown in the observed stream flows of the major tributaries and simulated runoff of the GCMs, is consistent with the increase in the frequencies of El Niño and La Niña events projected by recent studies [Cai, 2014] [Cai, 2015]. These changes in the frequencies of El Niño and La Niña events would change the Nile flow patterns following the observed tele-connection between ENSO and the Nile, in which El Niño and La Niña

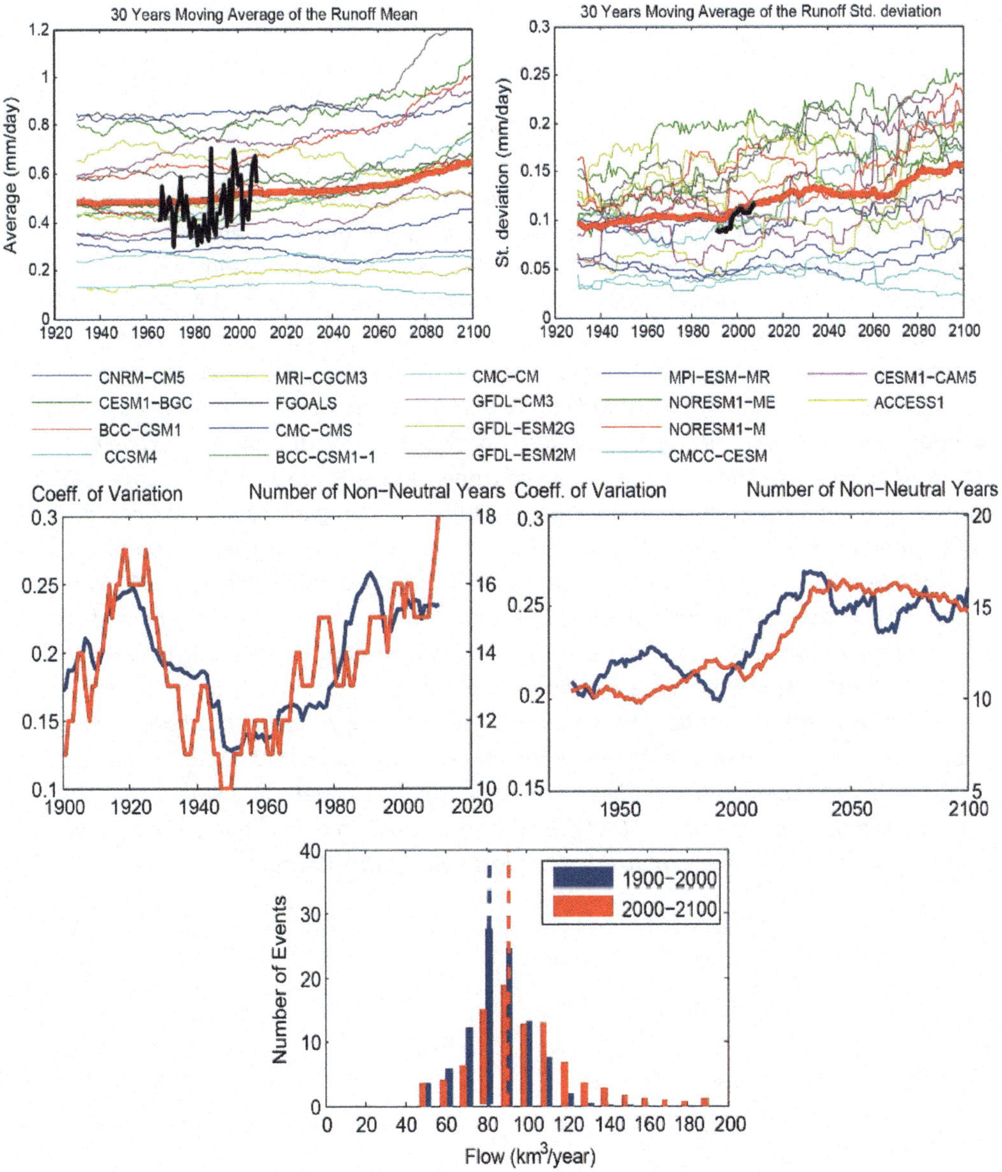
30 Years Moving Average of the Runoff Mean
Average (mm/day)
30 Years Moving Average of the Runoff Std. deviation
St. deviation (mm/day)
CNRM-CM5
MRI-CGCM3
CMC-CM
MPI-ESM-MR
CESM1-CAM5
CESM1-BGC
FGOALS
GFDL-CM3
NORESM1-ME
ACCESS1
BCC-CSM1
CMC-CMS
GFDL-ESM2G
NORESM1-M
CCSM4
BCC-CSM1-1
GFDL-ESM2M
CMCC-CESM
Coeff. of Variation
Number of Non-Neutral Years
Coeff. of Variation
Number of Non-Neutral Years
1900-2000
2000-2100
Number of Events
Flow (km³/year)

Figure 5.3 (opposite page): Changes in moving averages for the mean and standard deviation, coefficient of variation, number of ENSO events and frequency distribution of flow over Eastern Nile basin using 18 CMIP5 GCMs. a, b, 30-year moving averages of the mean and standard deviation of the runoff simulated by the models from 1900 to 2100, the 95% confidence levels of the mean and standard deviation using Student t and Chi-square tests are 0.54 (mm/day) and 0.11 (mm/day), respectively, based on the mean and standard deviation of the period (1900-1930). The thick black line is the observed annual flow of Atbara and Upper Blue Nile in a, and the 30-year moving average of the standard deviation of the flow of Atbara and Upper Blue Nile basin in b. The Sobat flows are not included as they were not available. c, 30-year moving average of the coefficient of variation (blue line) of the stream flow at Dongola with the number of non-neutral years (red line) (i.e. moderate and extreme El Niño and La Niña years). d, 30-year moving average of ensemble average of the coefficient of variation (blue line) of 18 GCMs of the runoff of the Eastern Nile basin with the number of non-neutral years (red line) (i.e. moderate and extreme El Niño and La Niña years). Changes in the frequency distribution in future Nile flow based on bias correction approach. The change in the long-term average of annual Nile flow for the future period (2000-2100; red dashed lines) is statically significant above the 95% confidence level using Student t-test compared the past period (1900-2000; blue solid lines). In a, b, c, and d the moving average is plotted at the last year of the 30 years.

events induce low and high flows in the river, respectively [Siam and Eltahir, 2015] [Wang and Eltahir, 1999] [ElDaw, et. al, 2003] (Figure 5.3d). It is also important to note that the increase in the variability in the flow of the Nile flow occurs between 2010 and 2040 and then stabilizes until the end of the 21st century, which is consistent with the weakening of the Walker circulation during the same period (2010-2040) [Kim, S. T. et al., 2014].

Although ENSO is a major driver of inter-annual variability in the Nile flow, and the tropical climate in general, this oscillation only explains 27% of the observed natural variability. In a different approach to the same problem, we use the ENSO-Nile tele-connection to predict future Nile flow based on changes in frequency of El Niño and La Niña events projected by an ensemble of GCMs. The future Nile flows estimated based on future El Niño and La Niña events exhibit the same changes in the mean, inter-annual variability and frequency distribution as the simulated runoff by the GCMs (See Sampling Analysis in Methods).

The agreement between the changes in the future flows projected using these two different approaches (i.e. simulated flow by the GCMs; and based on projected changes in El Niño and La Niña events) increases our confidence in the projected changes and support the mechanism proposed to explain the future changes in the Nile flows. Moreover, a sensitivity analysis of the results based on the choice of the GCMs using the sampling approach is conducted and the results are consistent regardless of which combination of models results is chosen to estimate the changes in the flow of the Nile. Our approach, which emphasizes the tele-connection between ENSO and the Nile flow, and our choice to use only GCMs that accurately simulate the statistics of ENSO resulted in narrowing the range of uncertainty about future Nile flow, and increasing our confidence in the projected impacts of climate change.

The relatively large increase in inter-annual variability in the flow of Nile river dictates an increase of almost 33% in the total storage capacity (from 340 to 460 Km^3) that is needed in order to maintain a flow in each year, for over 100 years, equivalent to the long-term mean flow (Figure 5.4a) (See Methods for details of the Hurst storage analysis [Hurst, et al., 1965] [Hurst, 1951]. It is also important to note that the long-term mean flow for the past (1900-2000) was about 80 km^3/year, while the mean flow for the future (2000-2100) is projected to be about 92 km^3/year. The current total storage capacity in the Eastern Nile river basin is about 155 km^3 and is expected to increase to 230 km^3 if the GERD is built (Figure 5.4a).

Based on our analysis shown in Figure 5.4a, applying the Hurst equation, the current storage in the basin is sufficient for securing a yield equivalent to the observed mean annual flow, for about 40 years. In order to secure a yield equivalent to the projected mean annual flow for a similar 40 years into the future, the storage will have to be expanded by about 55%. However, after GERD is built, the new storage in the basin will be sufficient for securing a yield equivalent to the observed flow, for about 60 years (Figure 5.4a). In order to secure a yield equivalent to the projected mean flow for a similar 60 years into the future, the storage will have to be expanded by about 45%

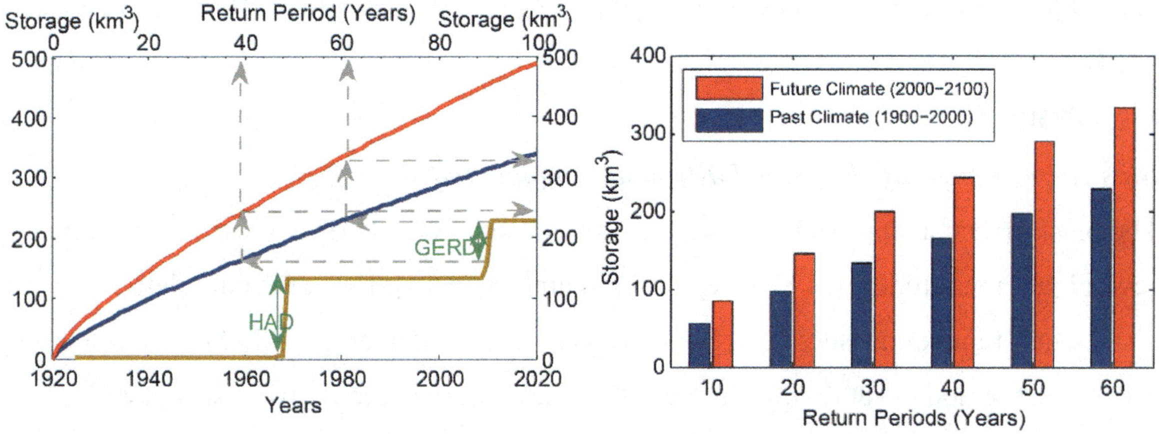

Figure 5.4: Total current water storage in the Eastern Nile basin, and required changes in future storage to accommodate the effects of climate change. a, The estimated water storage as function of the return period based on Hurst equation, the blue and red lines are for the past climate (1900-2000) and future climate (2000-2100) periods respectively (See Water Storage Analysis in Methods) and the increase in water storage capacity of the dams in the Eastern Nile basin with time including (Sennar, Khashm El Girba, Roseiras, High Aswan Dam (HAD), Fincha, Meroewe, Tekeze and Grand Ethiopian Renaissance Dam (GERD) to be completed by 2017). b, Water storage required for different return periods according to Hurst equation for the past climate (1900-2000) and future climate (2000-2100) periods.

(Figure 5.4b). It is important to note that our analysis does not include the evaporation losses from new reservoirs behind dams. Thus, we project that in the future water storage capacity in the basin will need to be reevaluated in order to control the enhanced inter-annual variability in the flow and to capture the additional flow due to climate change. However, this increase in storage capacity may or may not be feasible depending on several factors including availability of suitable sites for dams, economic considerations, and agreement between Nile basin countries. Moreover, detailed studies for these new dams that identify the potential locations, sizes, operational rules and impacts on the water resources shares between riparian countries based on a range of expected changes in the future flows of the Nile will be required [Cervigni, et al., 2015] [King and Block, 2014].

METHODS

Bias correction using the probability matching method.

The bias correction approach was used to correct for bias in hydrological variables in several previous studies [El.Shamy, 2009] [Yamana, 2015]. Here, the cumulative distribution function is calculated for the observed annual average Nile flow at Dongola for the past period (1900-2000). In the following step, the cumulative distribution function is calculated for the annual average simulated runoff by each GCM for the past period (1900-2000). Then, a set of bias correction factors are estimated by the ratio of the observed Nile flow at each cumulative probability to the corresponding Nile flow simulated by the GCM at that same cumulative probability. In other words, for each range of simulated Nile flow values produced by a GCM, we estimate a correction factor that would effectively match the observed and simulated probability distributions of past flows. For the period (2000-2100), the simulated runoff by each GCM is bias corrected by multiplying the GCM simulated value by the corresponding correction factor.

Sampling analysis

In the sampling approach, we estimate the Nile flow based on the frequency of occurrence of different types of El Niño and La Niña events (i.e., extreme El Niño, moderate El Niño, neutral, extreme La Niña and moderate La Niña). First, we compute the frequency of different events based on sea surface temperature (SST) observations using HadISST29 for the period (1900-2000). The identification process of the different types of events follows the methodology of previous studies that were investigating the changes in the frequencies of El Niño and La Niña events [Cai, 2014] [Cai, 2015].

In these studies, the extreme El Niño events are identified when the average (December-February) rainfall over Niño 3 region is greater than 5 mm/day. However, we replaced this condition by satisfying a SST condition over the Niño 3 region as there is no observed precipitation data available over the oceans for the whole period (1900-2000). It is estimated that when the rainfall over Niño 3 exceeds 5 mm/day, the average SST anomalies during December to February over Niño 3 are greater than 1.75 times the standard deviation. Thus, extreme El Niño events are identified when the (December-February) SST anomalies over Niño 3 are greater than 1.75 times the standard deviation. On the other hand, the extreme La Niña events are identified when the average (December-February) SST anomalies over Niño 4 are less than 1.75 times the standard deviation.

The moderate El Niño events are identified when the average (December-February) SST anomalies over Niño 3 are greater than 0.5 times the standard deviation and less than 1.75 times the standard deviation. The moderate La Niña events are identified when the average (December-February) SST anomalies over Niño 4 are less than 0.5 and greater than 1.75 times the standard deviation. The remainder events are considered as neutral events.

Second, for each type of event (i.e. extreme El Niño, moderate La Niña, neutral, extreme La Niña and moderate La Niña) within the past period (1900-2000), the corresponding observed total Nile flows at Dongola are identified and grouped together for each type.

The same identification process of the different types is repeated for GCMs simulations for the periods (1900-2000) and (2000-2100). The different types of events are identified for each model and for the past period (1900-2000). Each time the model simulates any event, a value is picked randomly from the group of the observed Nile flow values for this type of event and is assigned to be the corresponding Nile flow. This step is repeated for the future period (2000-2100), however, the magnitude of assigned flow is modified by the ratio of the simulated future runoff over ENB for that type of event in the period (2000-2100) to the past simulated runoff over ENB for the same type of event in the period (1900-2000).

Water Storage Analysis

The storage capacity required to accommodate the variability in any river flow and to supply an equal flow each year (i.e., the long term average) over a certain period of time (N) can be calculated by taking the departures of the flow from the mean and summing these departures [Wang and Eltahir, 1999] [El Daw, et al., 2003].

Thus, the difference between the maximum and minimum of these continued sums is the required storage capacity. Several phenomena which have long time series, including the Nile flow, follow a certain relation between the difference of maximum and minimum continued sum, the standard deviation (σ), and (N) [Kim, et al., 2014]. The relation has the form $R/\sigma = (N/2)k$, where k is the Hurst coefficient [Wang and Eltahir,

1999]. The Hurst coefficient can be estimated by fitting the relation between (R/σ) and (N/2) for different lengths of the time series of length (N).

Data availability

In this study, we use several observational data sets and climate models simulations to investigate the future change in the inter-annual variability of the flow in the Nile River. The observational data sets include stream flow data at Dongola available from the Global River Discharge Database (GRDD) [Vorosmarty, et. al, 1998], stream flow data for Atbara and Upper Blue Nile basins available through personal communications with the ministries of water in the countries of the Nile basin, and rainfall data-based rain gauges from the Global Historical Climatology Network (GHCN-V2) [Vose, et. al, 2016].

The climate models simulations are based on 18 GCMs that participated in the Coupled Model Inter-comparison Project Phase 5 (CMIP5) with the Representative Concentration Pathway (RCP 8.5) for future greenhouse gases projected emissions [Taylor, et. al, 2015] .

Business as Usual

The lack of consensus between the past studies that investigated the impacts of climate change on the Nile river flow hinders the development of effective climate change adaptation plans. Hence, knowledge about the future flow of the Nile river is needed to guide water resources planning including proposals for dams and hydropower projects (e.g. Grand Ethiopian Renaissance dam).

Based on our analysis using two different approaches and several different combinations of GCMs results, we project that the mean and inter-annual variability of annual flow in the Nile river basin will increase due to climate change. The projected increase in the inter-annual variability may limit the benefits of projected increase in mean flow and hence motivate a significant increase in the water storage capacity in order to fully utilize the additional water resource. Future climate projections discussed here assume a business-as-usual scenario of greenhouse gas emissions. Any effective mitigation efforts are likely to impact these projections and hence reduce the intensity of the projected impacts.

A version of this chapter first appeared as:

Siam, M. and Eltahir, E., 2017. Climate change enhances interannual variability of the Nile river flow. Nature Climate Change, 7(5), pp.350-354.

Chapter 6—Explaining and forecasting interannual variability in the flow of the Nile River

Mohamed S. Siam
Elfatih A. B. Eltahir

The Nile basin covers an area of $2.9x10^6$ km^2, which is approximately 10% of the African continent (Figure 6.1). It has two main tributaries, the White Nile and the Blue Nile, which originate from the equatorial lakes and Ethiopian highlands, respectively. The upper Blue Nile (UBN) basin is the main source of water for the Nile River. It contributes approximately 60% of the annual flow of the Nile and 80% of the total Nile flow that occurs between July and October at Dongola [Conway and Hulme, 1993] (Figure 6.2). The UBN basin extends over an area of $175x10^3$ km^2 (7° N to 12°5′ N and from 34°5′ E to 40° E). The mean annual rainfall over this basin is 1200 $mmyr^{-1}$ [Conway and Hulme, 1993]. Almost 60% of the annual rainfall over the UBN occurs during the summer between July and August, resulting in a largely predictable seasonal variability in the flow of the river.

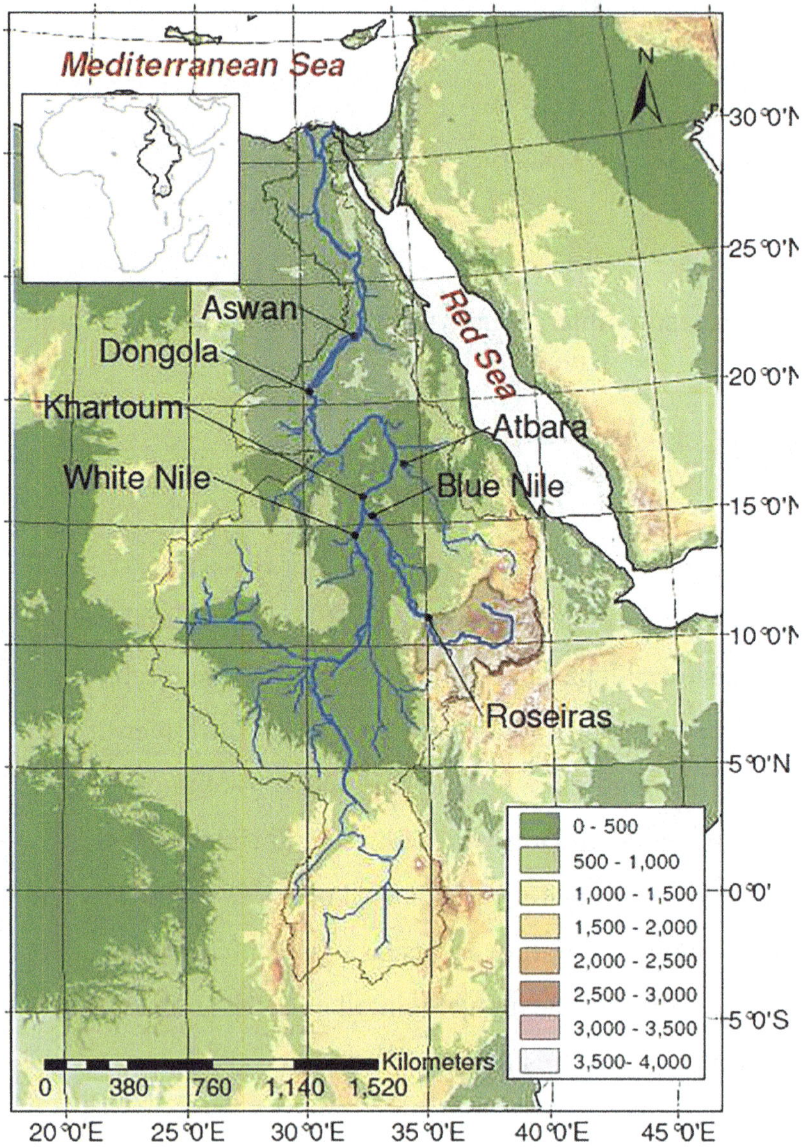

Figure 6.1: Topographic map of the Nile basin showing the outlet of the upper Blue Nile basin (shaded in gray) at Roseiras. The White and Blue Nile join together at Khartoum to form the main branch of the Nile that flows directly to Dongola in the north.

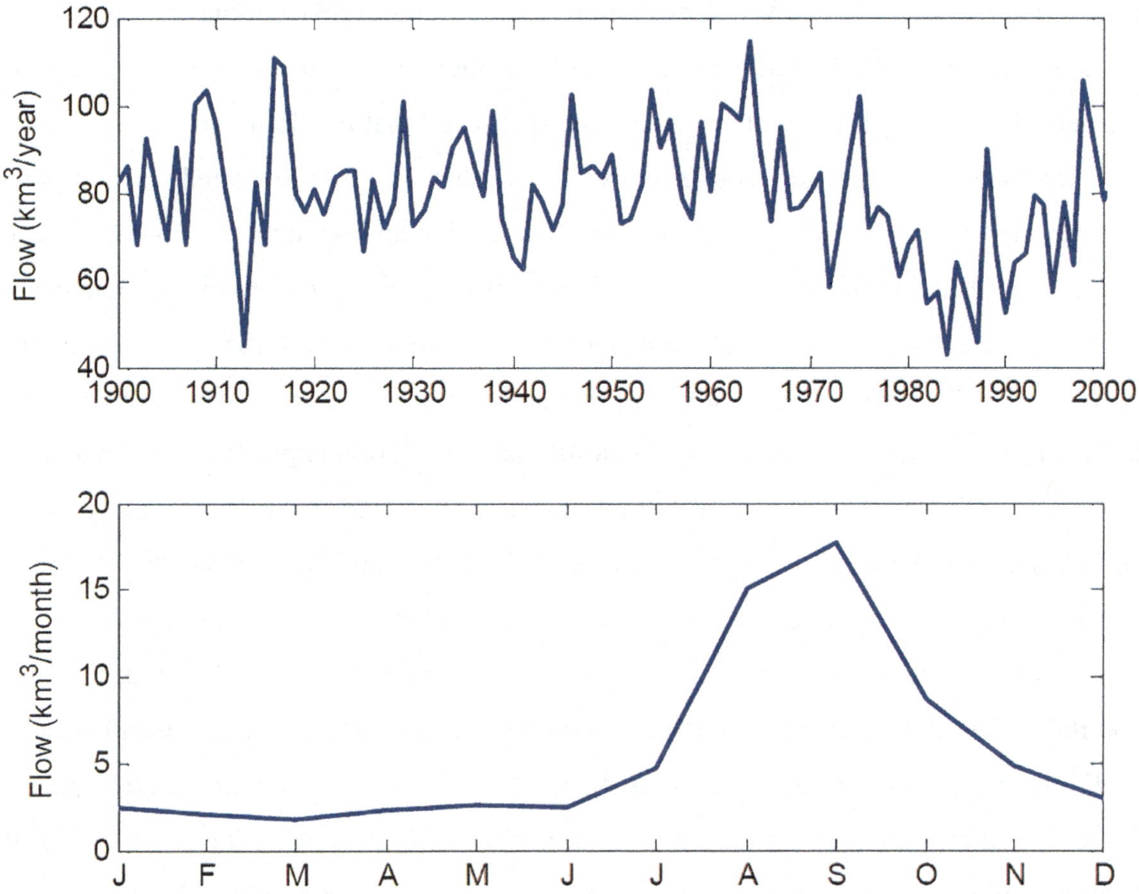

Figure 6.2: Annual Nile flow (top panel) and seasonal cycle (bottom panel) of the flow at Dongola for the period from 1900 to 2000.

The prediction of interannual variability in the flow of the Nile is rather challenging. Many studies, such as the one detailed in Chapter 2, investigated the teleconnections between the Ethiopian rainfall and the global sea surface temperature (SST) in order to find SST indices to use for Nile flow prediction [e.g., Abtew et al.,

2009; Melesse et al., 2011]. We showed in chapter 2 that the SST anomalies over the tropical eastern Pacific Ocean explain 25% of the interannual variability of Nile flow for the period 1872–1972. ElSanabary et al. [2014] showed that the dominant frequencies of the Ethiopian rainfall ranged from 2 to 8 years and that the scale-averaged wavelet power of the SSTs over the eastern Pacific and southern Indian and Atlantic oceans can explain a significant fraction of the rainfall variability over Ethiopia using wavelet principal component analysis. These correlations between the Nile flow and SST indices were the basis for new forecast models that were proposed to predict the Nile flows. For example, the work discussed in chapter 4 used a discriminant prediction approach to estimate the probabilities that the Nile flow will fall into prescribed categories. ElDaw et al. [2003] and Gissila et al. [2004] used SST over the Pacific, Indian and Atlantic oceans as predictors within a multiple linear regression model to predict the Nile flow.

The mechanisms behind these teleconnections between the rainfall over Ethiopia and the global SSTs were examined in several studies [e.g., Beltrando and Camberlin, 1993]. However, a clear distinction must be made between rainfall over the UBN basin in Ethiopia and rainfall over eastern Africa, defined as the region along the coast, east of the Ethiopian highlands (Figure 6.1). The UBN basin has one rainy season (May to September) during which more than 80% of the rainfall occurs, while along the eastern coast of Africa and depending on the location from the Equator, the seasonal cycle of rainfall can have two rainy seasons [Black et al., 2003; Hastenrath et al., 2011]. This pattern in the seasonal cycle of rainfall is related to the migration of the intertropical convergence zone (ITCZ) across the Equator. Camberlin [1995] showed that the rainfall over eastern Africa, including the UBN basin, is strongly coupled with the dynamics of the Indian monsoon. During strong Indian monsoon seasons, the sea level pressure over India decreases significantly, which enhances the pressure gradient between eastern Africa and India. As a result, westerly winds increase over eastern Africa and enhance transport moisture from the Congo basin to Ethiopia, Uganda and western Kenya. Giro et al. [2011] also showed that the warming over the Pacific Ocean, during El Niño events, reduces these westerly winds, which reduce the rainfall over eastern Africa. In addition, the monsoon circulation is weaker during El Niño events due to modulation of the Walker circulation and enhanced subsidence over the western Pacific and southern Asia; thus, the rainfall over Ethiopia decreases [Ju and Slingo, 1995; Kawamura, 1998; Shukla and Wallace, 1983; Soman and Slingo, 1997]. The reduced Nile flows during El Niño events were also attributed to the enhanced tropical-scale subsidence that suppresses rainfall, as a consequence of the increased upwelling over the eastern Pacific Ocean [Amarasekera et al., 1996].

The physical mechanism of the teleconnection between the Nile flow and SSTs of the northern and middle Indian Ocean and ENSO is described in another paper by the authors that will be explored in Chapter 7. Nile flow is strongly modulated by ENSO through ocean currents. During El Niño events, the warm water travels from the Pacific Ocean to the Indian Ocean through the Indonesian through flow and advection by the Indian Equatorial Current [Tomczak and Godfrey, 1995]. As a result, SSTs in the northern and middle Indian Ocean warm up following the warming of the tropical eastern Pacific, and force a gill-type circulation anomaly with enhanced westerly winds over the western Indian Ocean [Yang et al., 2007]. The latter enhances the low-level divergence of air and moisture away from the upper Blue Nile, resulting in a reduction of rainfall over the basin. On the other hand, the warming over the southern Indian Ocean generates a cyclonic flow in the boundary layer, which reduces the cross-equatorial meridional transport of air and moisture towards the UBN basin, favoring a reduction in rainfall and river flows. The teleconnections between the Pacific Ocean and the Nile basin and between the Indian Ocean and the Nile basin are reflected in different modes of observed natural variability in the flow of the Nile River, with important implications for the predictability of floods and droughts.

The objectives of this study were (i) to investigate the strength of the teleconnection between the Indian Ocean and the Nile basin and its role in explaining observed natural modes of variability in the flow of the Nile River, and (ii) to develop a new hybrid forecasting algorithm that can be used to predict the Nile flow based on indices of the SST in the eastern Pacific and southern Indian oceans.

Data

In this study, we used observed SSTs over the Indian and Pacific oceans from the monthly global (HadISST V1.1) data set on a 1° latitude–longitude grid from 1900 to 2000 [Rayner et al., 2003]. The monthly flows at Dongola from 1900 to 1984 were extracted from the Global River Discharge Database (RivDIS v1.1) [Vörösmarty et al., 1998] and from 1984 to 2000 through personal connection. The average monthly anomalies from September to November of the SSTs averaged over the eastern Pacific Ocean (6–2° N, 170–90° W; 2° N–6° S, 180–90° W; and 6–10° S, 150–110° W) are used

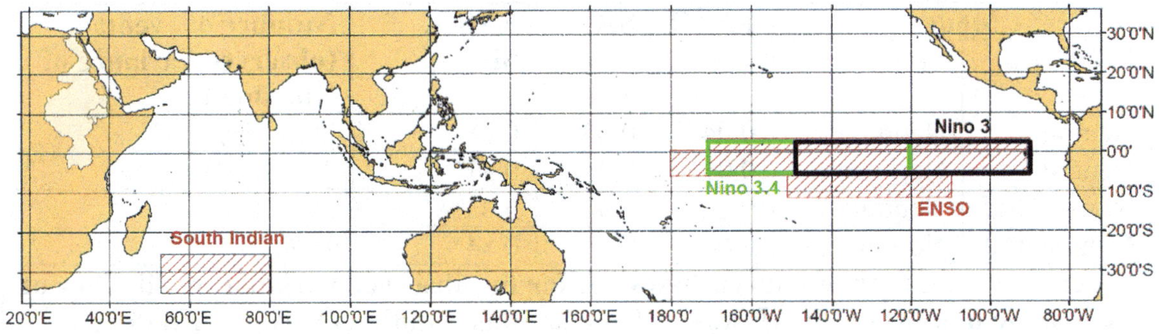

Figure 6.3: World map showing areas that cover the ENSO and northern and southern Indian Ocean SST indices. El Niño 3 and 3.4 are outlined in black and green, respectively. The whole Nile basin is outlined in black.

as an index of ENSO. This area has shown the highest correlation with the Nile flows, and it almost covers the same area as the Niño 3 and 3.4 indices [Trenberth, 1997].

Relation between the variability in the flow of the Nile River, ENSO and the Indian Ocean SST

Based on extensive correlation analysis of the Nile River flow at Dongola and the observed SST in the Indian Ocean, this study identified a region over the southern Indian Ocean (50– 80° E and 25–35° S) (see Figure 6.3) as the one with the highest correlation between SST and the Nile flow. This correlation is especially high for river flow (accumulated for July, August, September and October) and SST during the month of August. An earlier study by ElDaw et al. [2003] used SST indices over the Indian Ocean to predict the Nile flow; however, they focused on regions of the Indian Ocean that are different from the region that we used in defining the southern Indian Ocean (SIO) index. In other words, the region of the SIO was not used by ElDaw et al. [2003]. Table 6.2 describes the regions of the Indian Ocean identified in both studies. This study emphasizes that the proposed forecasting methodology for the Nile flow is motivated by

Mode		ENSO	SIO	ENSO, SIO	Number of events (observed variance of Nile flow)
ENSO	SIO				
Neutral	Neutral	0.04	0.03	0.08	29 (6.76)
Neutral	Non-neutral	0.05	0.28[a]	0.31[a]	26 (10.24)
Non-neutral	Neutral	0.1[a]	0.02	0.43[a]	26 (5.8)
Non-neutral	Non-neutral	0.64[a]	0.6[a]	0.84[a]	19 (12.3)

Note: the values between parentheses are for the observed variance of Nile flow of each mode in units of MCM^2 day^{-2}. SIO: southern Indian Ocean SST index; ENSO: ENSO index. [a] Values that are significant at the 1% significance level.

Table 6.1: Summary of the coefficient of determination (R^2) between the average Nile flow from July to October and different combinations of indices of ENSO and SIO.

Region	Location	Study
1	(35–44° S, 115–130° E)	ElDaw et al. (2003)
2	(0–7° S, 90–130° E)	
3	(35–44 S, 20–60° E)	
4	(10–20° S, 110–125° E)	
5	(50–80° S, 25–35° E)	This study

Table 6.2: Comparison between regions in the Indian Ocean used in ElDaw et al. (2003) and this study to predict the Nile flow.

the physical mechanisms proposed by Siam et al. in Chapter 7 and described earlier in this chapter. In contrast, the forecasting approach of some of the previous studies was based on purely statistical correlations found between the Nile flow and SSTs globally.

Figure 6.4 shows the observed and simulated time series of the average July-to-October Nile flow at Dongola, which accounts for approximately 70% of the annual Nile flow. The Nile flow is predicted by three different linear regression models using either ENSO averaged from September to November (Figure 6.4a) or SIO August (Figure 6.4b) indices, or both (Figure 6.4c) as covariates. It is clear from this figure that the addition of the SIO index increases the explained variability of the Nile flow to 44 %, compared to only 30% when the ENSO index is used alone. This indicates that the SIO index can explain almost 14% of the variability of the Nile flow that is independent of ENSO. The north and middle of the Indian Ocean have also exhibited a high correlation between their SST and the Nile flow. However, the additional variability explained by the SST over the northern and middle Indian Ocean, when combined with the ENSO index, is negligible (not shown here). This is mainly because the SSTs over the northern and middle Indian Ocean are dependent on ENSO, while the SST over the southern Indian Ocean (i.e., the SIO index) is not.

In further analysis, we define ±0.5° C as the threshold between non-neutral and neutral years in the eastern Pacific Ocean, based on the ENSO index. This value is about two thirds of 1 standard deviation of the anomalies of the ENSO index. The same threshold has been used to identify nonneutral and neutral years using the El Niño 3.4 index, which is similar to our ENSO index [Trenberth, 1997]. This indicates that if the ENSO index anomaly is greater than 0.5 C or less than −0.5 C, it is considered a non-neutral condition; otherwise, it is considered a neutral condition. Similarly, a ±0.3 C value is used as a threshold between non-neutral and neutral years in the southern Indian Ocean using the SIO index. This value is also about two-thirds of 1 standard deviation for the anomalies of the SSTs over this region. Thus, if both ENSO and SIO indices are used together, four different combinations can be defined based on these classifications. The first is when both ENSO and SIO indices are neutral (29 out of 100 events), the second is

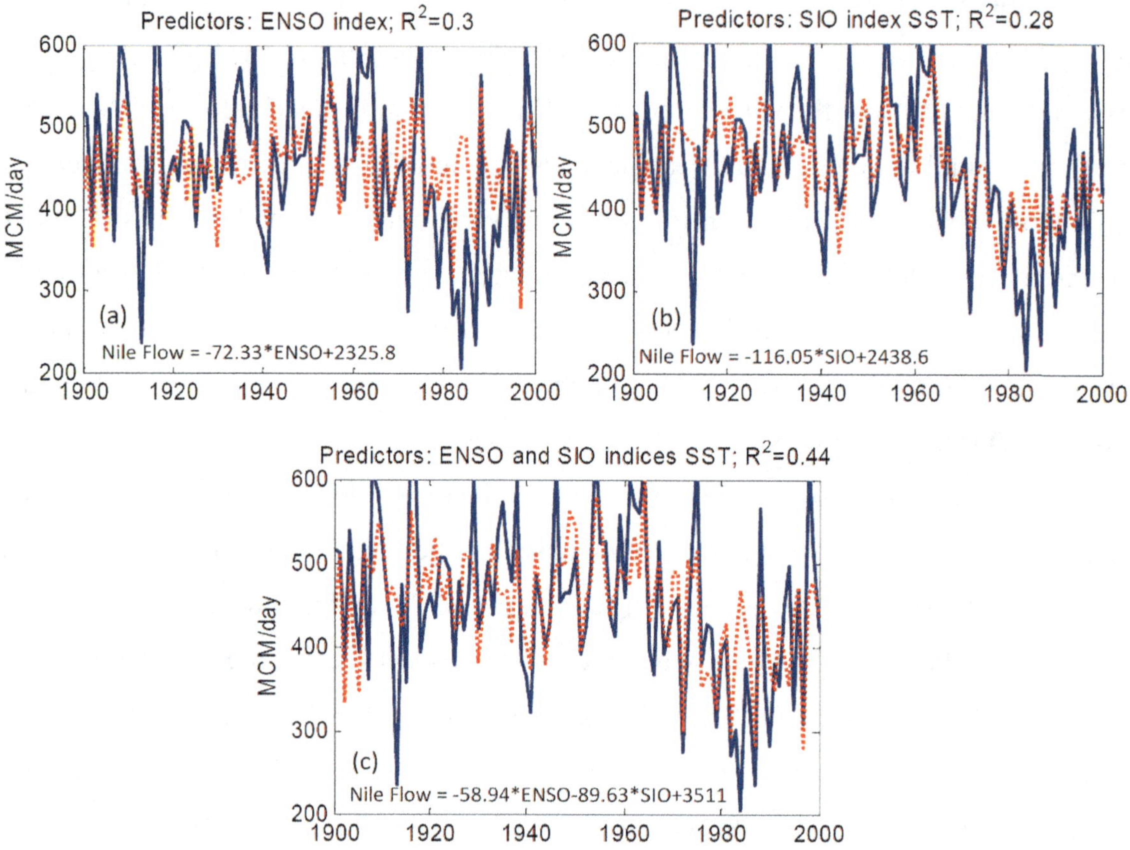

Figure 6.4: Observed (solid blue lines) and simulated (dashed red lines) average Nile flows in million cubic meters per day (MCM day^{-1}) from July to October at Dongola using the (a) ENSO index, the (b) SIO index and the (c) ENSO and SIO indices as predictors for the period 1900 to 2000. The simulated flows are calculated based on the equations shown in each figure, in which the predictands are the average observed SSTs over the ENSO and SIO regions in degrees Celsius and the predictor is the average Nile flow from July to October in million cubic meters per day (MCM day^{-1}).

when both ENSO and SIO indices are non-neutral (19 out of 100 events), the third when SIO is non-neutral and ENSO is neutral (26 out of 100 events) and, finally, when SIO is neutral and ENSO is nonneutral (26 out of 100 events). Each of these combinations is considered a mode of natural variability in the flow of the Nile River. Then, the Nile flow is calculated as a predictand using multiple linear regression, with the ENSO and SIO indices of each mode as predictors.

Four different modes are identified for describing the natural variability in the flow of the Nile River and are summarized in Table 6.1. The ENSO and SIO indices do not explain a significant fraction of the interannual variability in the flow of river when they are both neutral (Figure 6.5a). The variability of the Nile flow in such years can be regarded as a reflection of the chaotic interactions between the biosphere and atmosphere and within each of the two domains. For this mode, the predictability of the Nile flow is rather limited. The other two intermediate modes include non-neutral conditions in the eastern Pacific and neutral conditions in the southern Indian Ocean, or vice versa (Figure 6.5b and c). For these two modes, a significant fraction (i.e., 31 and 43 %) of the variance describing interannual variability in the flow is explained. Hence, these modes point to a significant potential for predictability of the flow.

Finally, indices of ENSO and SIO can explain 84% of the interannual variability in the Nile flow when non-neutral conditions are observed for both the eastern Pacific and southern Indian oceans (Figure 6.5d). Therefore, the SIO index can be used to predict the flow together with the ENSO index, as collectively they can explain a significant fraction of the variability in the flow of the Nile River. This result indicates that, during

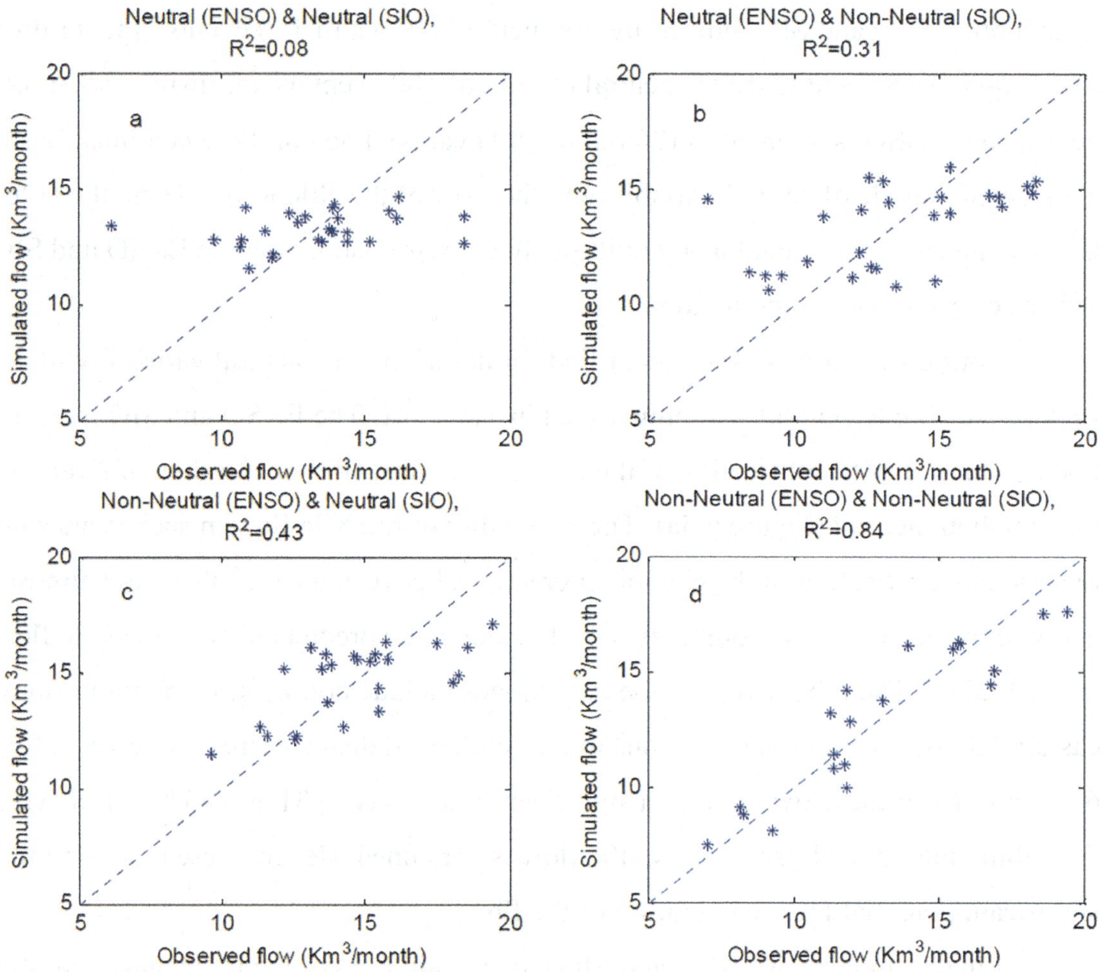

Figure 6.5: A comparison between the observed and simulated Nile flow showing the different modes of variability for the period from 1900 to 2000: (a) neutral ENSO (29 events) and SIO, (b) neutral ENSO and non-neutral SSTs in SIO (26 events), (c) non-neutral ENSO and neutral SSTs in SIO (26 events) and, finally, (d) non-neutral ENSO and non-neutral SSTs in SIO (19 events).

years with anomalous SST conditions in both oceans, floods and droughts in the Nile River flow can be highly predictable, assuming accurate forecasts of those indices are available.

A hybrid methodology for long-range prediction of the Nile flow

We proposed a simple methodology to predict the Nile flow with a lead time of about a few months (3–6 months). The forecast of global SST distribution based on dynamical models [e.g., NCEP coupled forecast system model version 2 (CFSv2); Saha et al., 2010, 2013] can be used together with the algorithm developed in this study to relate the Nile flow to the ENSO and SIO indices. The proposed method is shown in Figure 6.6 and can be described in two main steps.

- Forecast of SST anomalies in the Indian Ocean and eastern Pacific Ocean using dynamical models of the coupled global ocean atmosphere system. Such forecasts are routinely issued by centers such as NCEP and ECMWF.
- Application of a forecast algorithm between the Nile flow (predictand) and forecasted SSTs in the Indian and eastern Pacific oceans (predictors) for the identified mode of variability.

In this chapter, we focused on the second step of the proposed method: the development of the algorithm relating SSTs and the Nile flow. We developed the forecast algorithm using observed SSTs. We did not describe how this algorithm can be applied with forecasts of global SST distribution based on dynamical models, as this step is beyond the scope of this chapter. However, we recognize that the overall accuracy of this method in predicting interannual variability of the Nile flow is dependent on the skill of

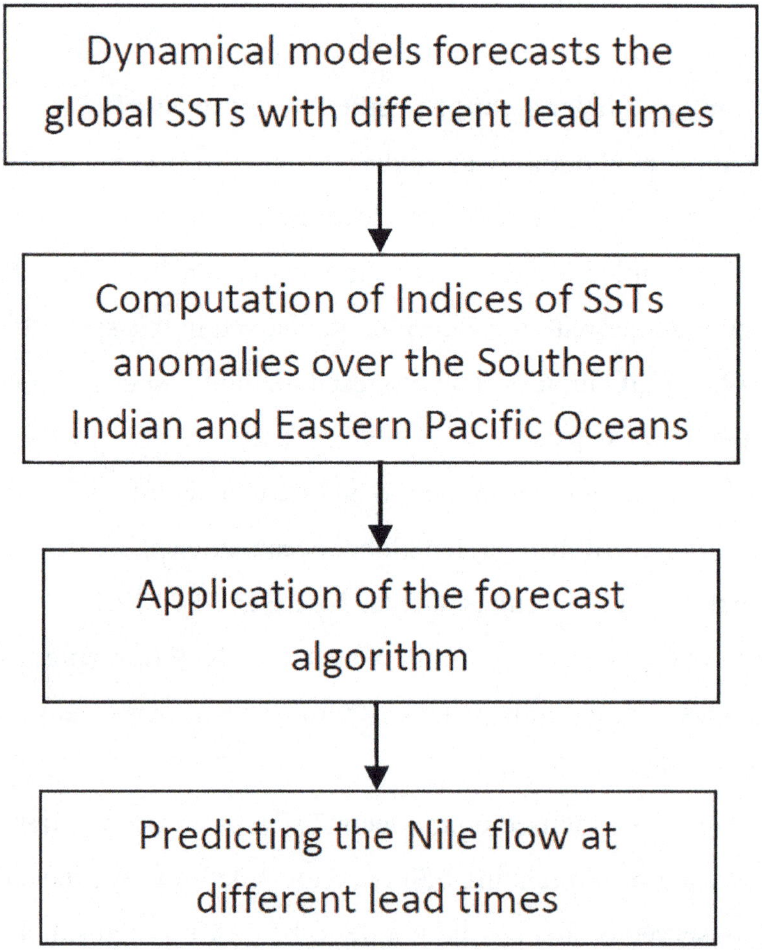

Figure 6.6: Schematic of the hybrid methodology for predicting the Nile flow using the SST forecasts of the dynamical models and the proposed forecast algorithm.

global coupled models in forecasting the global SSTs. Thus, the selection of the forecast model that predicts the SSTs is an important step to ensure the accuracy of the prediction of the Nile flow. As global coupled ocean–atmosphere models improve in their skill of forecasting global SSTs in the Pacific and Indian oceans, we expect that our ability to predict the interannual variability in the Nile flow will improve too. In addition, the accuracy in the prediction of the Nile flow on medium and short timescales (of weeks to 1 month) can be improved by adding other hydrological variables (e.g., rainfall and stream flow) over the basin, as demonstrated by the research discussed in Chapter 4. The proposed method can be described as hybrid since it combines dynamical forecasts of global SSTs and statistical algorithms relating the Nile flow and the forecasted SSTs. The same method can also be described as hybrid since it combines information about SSTs from the Pacific and Indian oceans.

In this research, we applied a discriminant approach that specifies the categoric probabilities of the predictand (Nile flow) according to the categories that the predictors (i.e., the ENSO and SIO indices) fall into. As noted in previous chapters, the annual Nile flow is divided into "low", "normal", and "high" categories. The boundaries of these categories are defined so that the number of points in each category is about one-third that of the data points (Figure 6.7). On the other hand, the ENSO and SIO indices are divided into "cold", "normal" and "warm" categories. (The words "normal" and "neutral" are used to describe the same conditions.) The boundaries for the normal category are −0.5° and 0.5° C for the ENSO index and −0.3° and 0.3° C for the SIO index (Figure

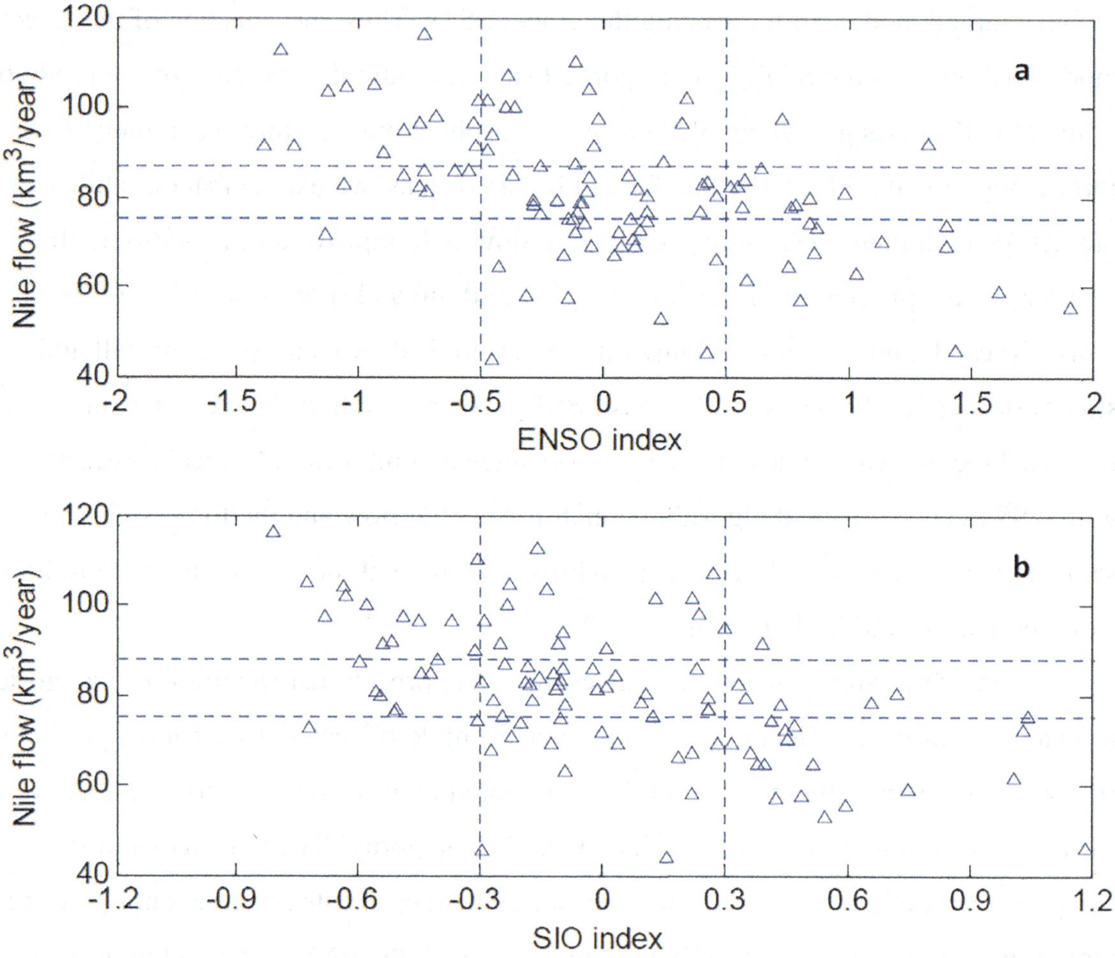

Figure 6.7: Relations between the annual Nile flow and different indices for the period (1900–2000): (a) ENSO and (b) SIO. The horizontal lines represent the boundaries for the "high", "normal" and "low" categories of the annual flow. The vertical lines represent the boundaries for the "warm", "normal", and "cold" conditions for the ENSO and SIO indices.

6.7). Any condition below the lower limit is considered "cold", and higher than the upper limit is considered "warm" for both indices.

The Bayesian theorem, described in Chapter 4 and in many statistical books [e.g., Winkler 1972; West, 1989], states that the probability of occurrence of a specified flow category (Q_i) and given two conditions (A and B) can be expressed as

$$P(Q_i/A,B) = \frac{P(B/Q_i,A)P(Q_i/A)}{P(B/A)} \qquad (6.1)$$

where $P(Q_i/A)$ is the probability of event Q_i given that event A has occurred, and $P(Q_i/A,B)$ is the probability of event Q_i given that events A and B have occurred, and similarly for other shown probabilities. In addition, if the events A and B are independent, we can rewrite Equation (6.1) as

$$P(Q_i/A,B) = \frac{P(B/Q_i)P(Q_i/A)}{\sum_{i=1}^{3} P(B/Q_i)P(Q_i/A)} \qquad (6.2)$$

The advantage of assuming independence between A and B and using Equation (6.2) is that it simplifies the calculation of $P(B/Q_i,A)$, since we do not have to split the data into a relatively large number of categories, which reduces the error due to the limitation of the data size (i.e. 100 years of data). The independence between the ENSO and SIO indices is a reasonable assumption, as the coefficient of determination between them is less than 6%. In order to evaluate the predictions of the Nile flow, we use the forecasting index (FI) defined Chapter 4 as

$$\mathrm{FP}(j) = \sum_{i=1}^{3} P_r(i,j)P_p(i,j) \qquad (6.3)$$

$$\mathrm{FI} = \frac{1}{n}\sum_{j=1}^{n} \mathrm{FP}(j) \qquad (6.4)$$

where $\mathrm{FP}(j)$ is the forecast probability in a certain year (j) and the FI is the average of the FP over a certain period, n. The prior probability $P_r(i, j)$ is calculated using Equation (6.2) for a certain year (j) and category (i = 1, 2, 3) and the posterior probability $P_p(i, j)$ is defined as [1, 0, 0] in a low flow year, [0, 1, 0] in a normal year, and [0, 0, 1] in a high flow year. Hence, a larger FI indicates a higher accuracy of the forecast. The FI without any information about SST should be about one-third, as we have classified flow data into three categories, each with a similar number of data points.

The data were split into a calibration period (1900–1970) and a verification period (1970–2000). Tables 6.3 and 6.4 summarize the conditional probabilities of Nile flow given certain conditions of the SIO or ENSO index. It is shown that during warm and cold conditions of SIO, the probabilities are significantly higher for low and high Nile flow, respectively. The same is true for the ENSO, as was described in Chapter 2. Table

SIO	**High**	**Nile flow Normal**	**Low**
Warm	0	0.25	0.75
Normal	0.23	0.39	0.39
Cold	0.57	0.26	0.17

Table 6.3 (above): Conditional probability of the Nile flow given SIO conditions.
Table 6.4 (below): Conditional probability of the Nile flow given ENSO conditions.

ENSO	**High**	**Nile flow Normal**	**Low**
Warm	0.15	0.31	0.54
Normal	0.22	0.38	0.41
Cold	0.68	0.32	0

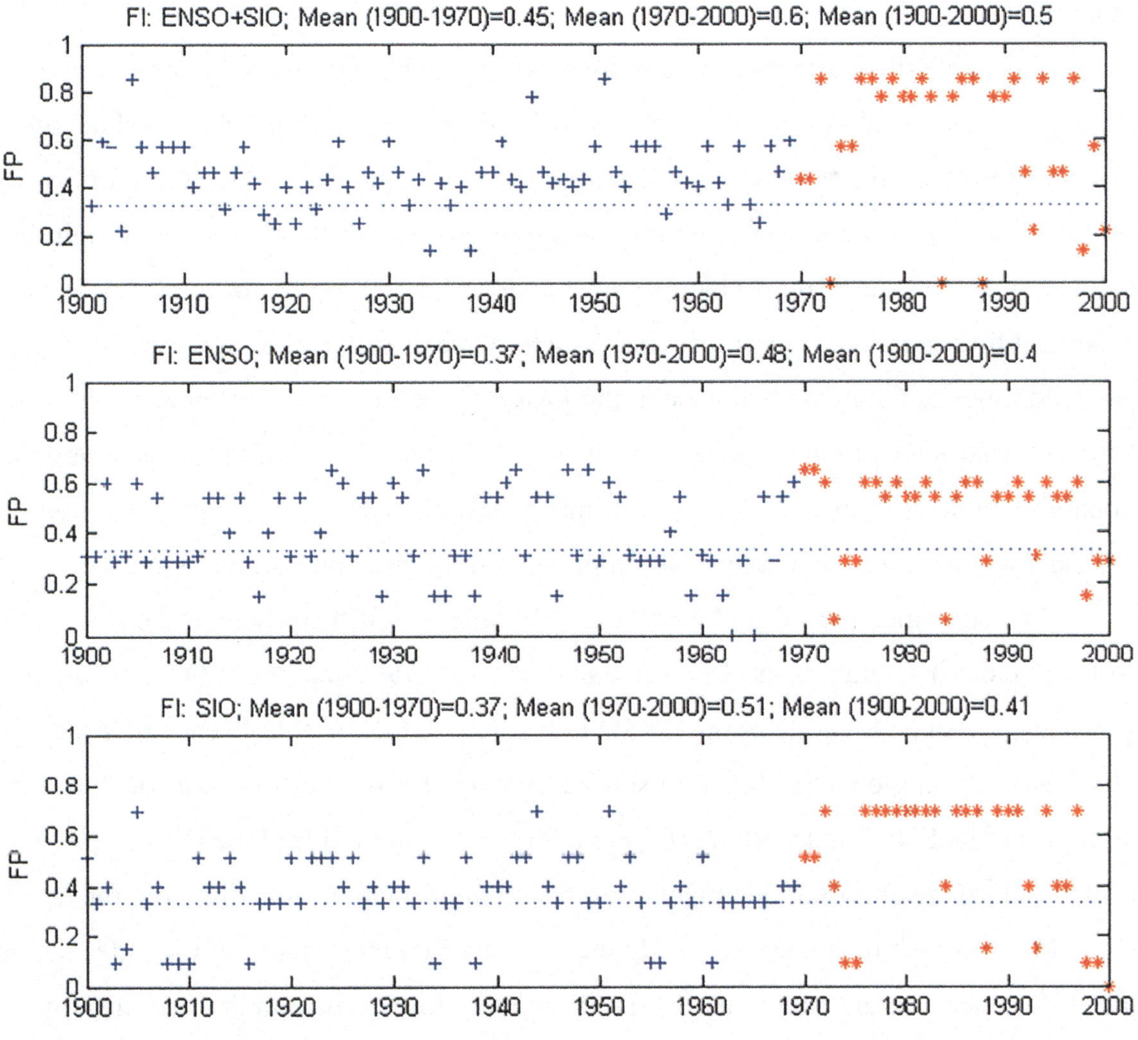

Figure 6.8: Time series of the forecast probability using different indices: (a) ENSO and SIO together, (b) ENSO, and (c) SIO. The period (1900–1970) is used for calculating the probabilities (shown in crosses) using Eq. (2) and 1970–2000 for validation (shown in stars).

6.5 shows the probabilities that are conditioned on both SIO and ENSO, calculated using Equation (6.2).

This table illustrates clearly how forecasts of the Nile flow can be improved by combining the two indices. For example, warm conditions in both oceans translate into 85% probability of low flow in the Nile, and an insignificant probability of high flow. On the other hand, cold conditions in both oceans translate into 83% probability of high flow in the Nile, and insignificant probability of low flow. Depending on the accuracy of the dynamical forecast models of global SSTs, such forecasts of the Nile flow can be issued with lead times of 6 months. At present, the Eastern Nile Regional Technical Office (ENTRO) issues operational forecasts of the Nile flow based on ENSO forecasts and the probability table described in Chapter 2 (similar to Table 6.4). We anticipate that use of Table 6.5 would represent a significant improvement in these operational forecasts.

The combined use of the ENSO and SIO indices significantly increased the FI to 0.5 (Figure 6.8a). Comparison of Figure 6.8b and c illustrates that the SIO index alone has almost the same FI value as the ENSO index. Recall that, in the absence of any information about global SSTs, the FI should have a value of one-third. The deviations of the FI using the ENSO index alone (Figure 6.8b) or the SIO index alone (Figure 6.8c) from one-third are almost added together to create the deviation of the FI from the hybrid method from one-third (Figure 6.8a). Hence, the new SIO index plays a role independent of ENSO in shaping the interannual variability in the flow of the Nile River. Thus, by using these two indices, we explained a significant fraction of the interannual variability

SIO	Nile flow	ENSO		
		Warm	Normal	Cold
SIO warm	High	0	0	0
	Normal	0.15	0.22	1
	Low	0.85	0.78	0
SIO normal	High	0.1	0.14	0.57
	Normal	0.31	0.4	0.43
	Low	0.59	0.46	0
SIO cold	High	0.33	0.42	0.83
	Normal	0.29	0.33	0.17
	Low	0.37	0.25	0

Table 6.5: Conditional probability of the Nile flow given SIO and ENS conditions.

in the flow of the Nile River, and illustrated a significant potential for improving the Nile flow forecasts.

A version of this paper first appeared as:

Siam, M. and Eltahir, E., 2015. Explaining and forecasting interannual variability in the flow of the Nile River. Hydrology and Earth System Sciences, 19(3), pp.1181-1192.

Chapter 7—Role of the Indian Ocean sea surface temperature in shaping the natural variability in the flow of Nile River

Mohamed S. Siam
Guiling Wang
Elfatih A. B. Eltahir

As discussed in previous chapters, the inter-annual variability in the flow of the Nile River is strongly correlated with El Niño Southern Oscillations (ENSO), its warm phase being associated with a reduced Nile flow. However, the mechanism behind this teleconnection is not yet fully understood.

The connection between El Niño events and the rainfall over East Africa was investigated in several studies [e.g. Beltrando and Camberlin 1993; Black et al. 2003]. However, a clear distinction must be made between rainfall over the UBN basin (Figure 7.1) and rainfall over East Africa, which is mainly along the East coast and Horn of Africa and away from the Ethiopian highlands. The mechanism that connects rainfall over these regions to ENSO is different from the corresponding mechanism in the UBN

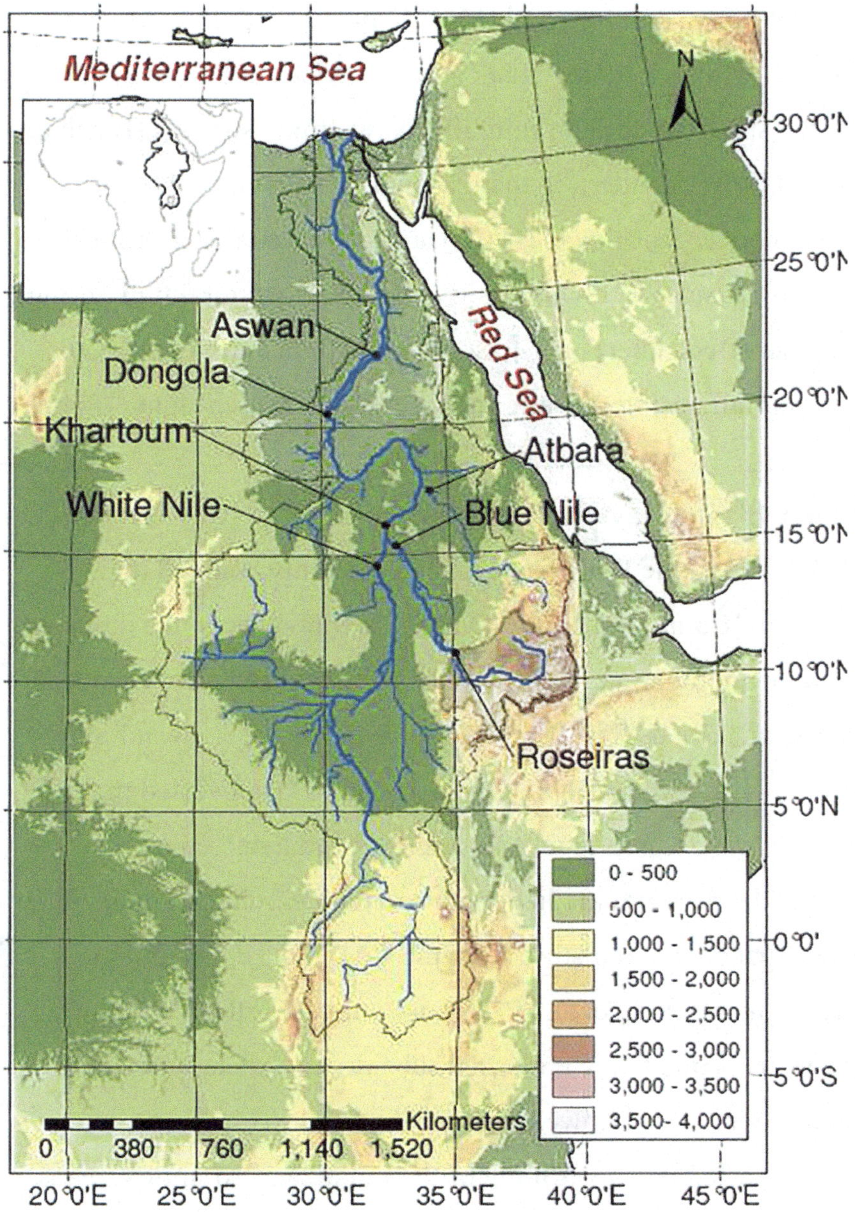

Figure 7.1: Topographic map of the Nile basin showing the outlet of the UBN basin (shaded in gray) at Roseiras. The White and Blue Nile Rivers meet at Khartoum to form the main branch of the Nile River that flows directly to Dongola in the North.

basin. The UBN basin has one rainy season period (May–September) during which more than 80% of its rainfall occurs, while along the East coast of Africa, and depending on the location from the equator, the seasonal cycle of rainfall has two rainy seasons [Black et al. 2003; Hastenrath et al. 2011]. This pattern in the seasonal cycle is related to the migration of the intertropical convergence zone (ITCZ) across the equator over these regions. In addition, the warm phase of ENSO is positively correlated to rainfall during the short rainy season (October–December) over East Africa [Mutai and Ward 2000], while it is not correlated to rainfall during the long rainy season in March–June; [Pohl and Camberlin 2006].

The rainfall over the UBN in (May–September) season is negatively correlated to ENSO Rainfall over East Africa and the UBN basin are significantly coupled to the intensity of the Indian monsoon [Camberlin 1995]. During El Niño events, the monsoon circulation is weaker due to the modulation of the Walker circulation and the enhanced subsidence over West Pacific and South Asia [Shukla and Wallace 1983; Ju and Slingo 1995; Soman and Slingo 1997; Kawamura 1998]. Camberlin [1997] suggested that during strong Indian monsoon seasons, sea level pressure over India decreases significantly, which enhances the pressure gradient between East Africa and India, thus increases the westerly winds that advect moisture from the Congo basin to Ethiopia, Uganda and western Kenya. Amarasekera et al. [1996] also suggested that the reduced Nile flow during El Niño events is due to the enhanced subsidence that suppresses rainfall as a consequence of the increased upwelling over the Eastern Pacific Ocean. Giannini et al. [2003] also suggested that a warming of the Indian and Atlantic Oceans leads to a reduction in rainfall over the Sahel region by enhancing convection over ocean and by decreasing convergence of air over land.

The study discussed in this chapter had two aims:

(1) we investigated the connection between the Nile flow and ENSO, and in particular, the role of Indian Ocean (IO) SSTs in shaping this teleconnection;

(2) we developed new indices, based on SSTs over Eastern Pacific and IOs, intended to improve long-term and midterm climate prediction of the Nile River flow.

Description of data used in this study

The data used in this study consisted of observations of river flow and global SST distribution together with reanalysis products that provide information on the atmospheric circulation.

The longest recorded and most accurate hydrological variable that we considered in this study was the measured stream flow at Dongola (Figure 7.1), which can be used as an index of rainfall over the UBN in Ethiopia. The flow at Dongola includes contributions from both the White and UBN rivers, and is referred hereafter as the Nile flow. However, the signature of the UBN flow on the Nile flow is dominant. The UBN provides more than 60% of the annual mean Nile flow at Dongola. Moreover, more than 80% of the JJASO (June–October) mean Nile flow is a result of the summer rainfall over the UBN. More importantly, the flow in the Blue Nile explains most of the inter-annual variability in the flow of the Nile River. The monthly mean flow data at Dongola were extracted from the Global River Discharge Database (RivDIS v 1.1) for the period 1871–2000 [Vörösmarty et al. 1998].

The SST data were extracted from the global monthly mean (HadISST V 1.1) dataset, available on a 1° latitude– longitude grid from 1871 to 2000 [Rayner et al. 2003]. Monthly anomalies of the SSTs were averaged over three Eastern Pacific Ocean regions (2°N–6°N, 170°W–90°W; 6°S–2°N, 180°W–90°W; 10°S–6°S, 150°W–110°W) and used as ENSO indices. These regions have the highest correlations with the mean Nile flow and cover the same areas as the Niño 3 and 3.4 indices [Trenberth 1997]. In addition, monthly anomalies of SSTs were averaged over three different regions in the IO. These regions are defined hereafter as the North IO (0–15°N, 50°E–70°E), the Middle IO (10°S– 0, 50°E–70°E) and the South IO (25°S–35°S, 50°E–80°E). Also, these regions exhibit the highest correlations with the Nile flow, compared to other regions of the IO. The variables considered to study the atmospheric circulation in response to changes in SST include zonal and meridional wind speed components on pressure levels. These were obtained from the ERA-Interim (ERAI) reanalysis product for the period 1979–2010 [Dee et al. 2011]. ERAI was chosen among other reanalysis products because it has the most accurate representation of the hydrological cycle over the UBN basin [Siam et al. 2013] and wind circulation over Africa compared to other reanalysis products such as ERA40 [Uppala et al. 2005] or NCEP–NCAR [Kalnay et al. 1996].

Observational analyses of the relationships between Nile flow and SSTs in Indian and Pacific Oceans

Cross-correlation analyses

In this section, we present a detailed analysis of correlations between SSTs over the Indian and Pacific Oceans and the Nile flow. From July to October, tropical Eastern Pacific SSTs show a persistent and significant negative correlation with the JASO (July–October) mean Nile flow for the period (1871–2000) as illustrated in Figure 7.2a–d. Figure 7.3a–d show the correlation between SON (September–November) mean ENSO index and the global SST distributions in July, August–September, and October. The comparison between Figs. 2 and 3 reveals that the correlation between the Nile flow and the global SSTs seems to reflect the correlation between ENSO and those SSTs. Therefore, the global SSTs in regions other than the East tropical Pacific may not explain any significant additional variability in the Nile River flow. One significant exception is the correlation between JASO mean Nile flow and South IO SSTs in August. This correlation is significant ($R^2 \sim 0.3$) and is not simply a reflection of the significantly smaller correlation ($R^2 \sim 0.2$) between ENSO index and SSTs in the same region of the South IO, and in the same month of August. This result suggested that the South IO may play an independent role from ENSO in shaping the variability of the Nile flow.

In order to estimate the highest correlation between the Nile flow and ENSO, Figure 7.4 shows the correlation between the ENSO index, considered separately for each month of the year, and the mean Nile flow over several months of the rainy season (JJAS: June–September, JASO: July–October, ASON: August–November). The JASO mean

Nile flow has the highest negative correlation with ENSO index, especially during the months of September–November. This result may be explained as due to the fact that all the Nile flow from July to October is necessary to consider in order to capture all the variability in the rainfall over Ethiopia during the entire rainy season from June to

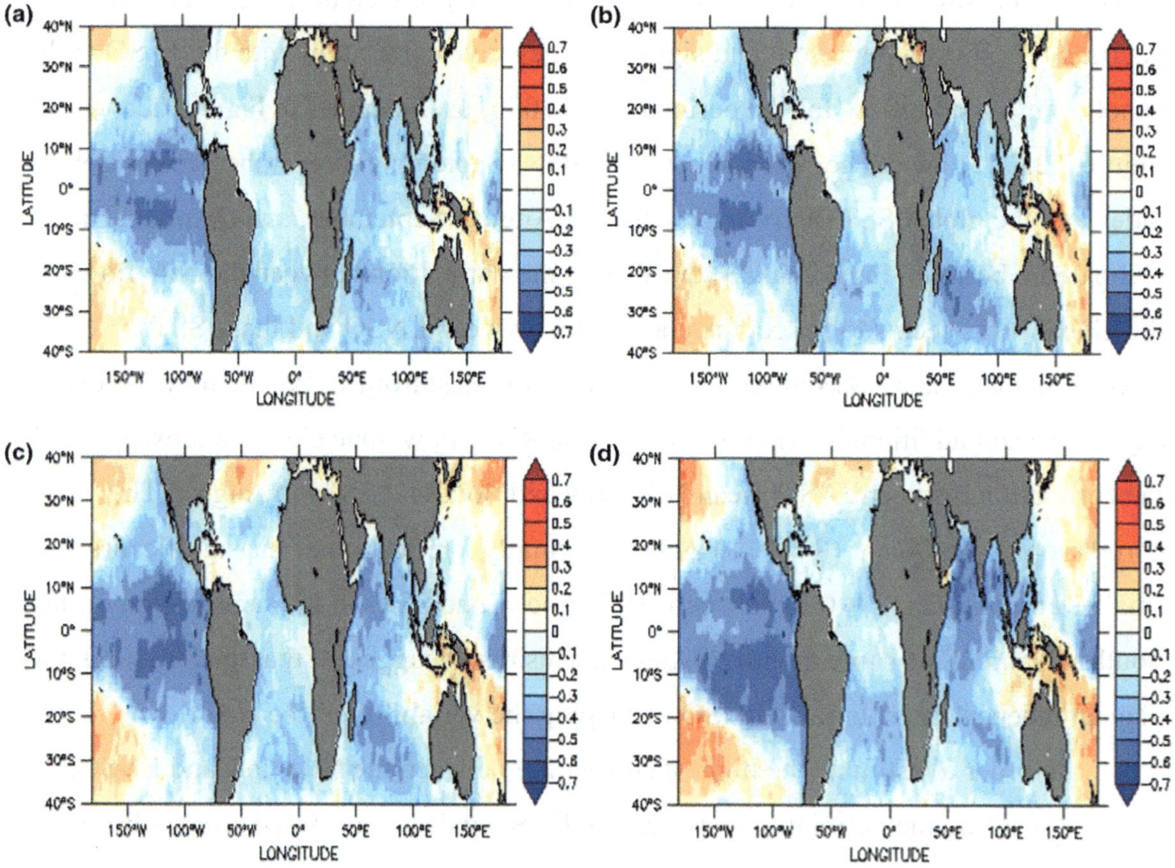

Figure 7.2: Correlations for the period (1871–2000) between average Nile flow during (July–October) and average monthly SST: a July, b August, c September and d October. The 1 % significance level of the correlation is 0.23

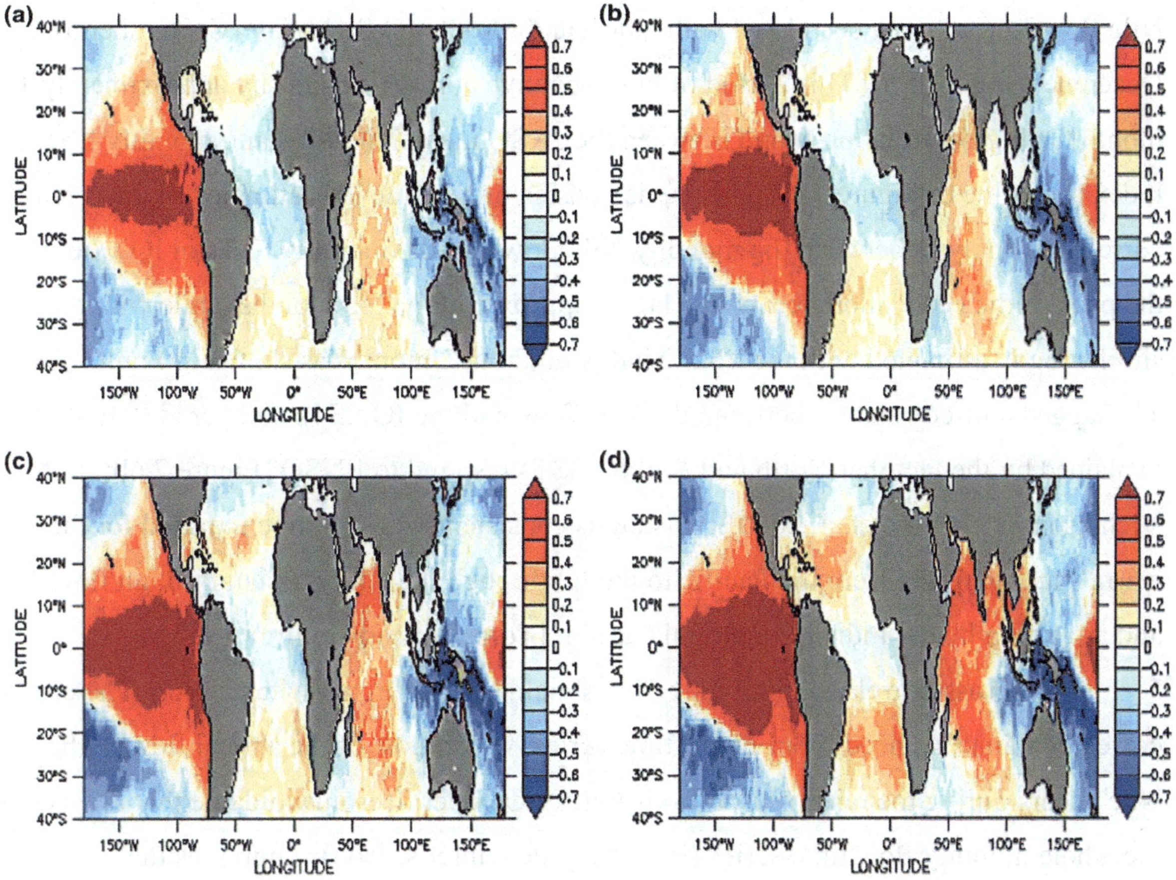

Figure 7.3: Correlations for the period (1871–2000) between average ENSO index during (September–November) and average monthly SST: a July, b August, c September and d October. The 1 % significance level of the correlation is 0.23

September and the signal of ENSO is the strongest during September– November (as discussed in Chapter 4).

The correlation analysis is repeated between the Nile flow averaged for JASO and monthly SSTs averaged over three different regions in the IO: North IO, Middle IO, South IO. The SST is considered separately for each month of the year. The JASO mean

Nile flow is negatively correlated with North and Middle IO SSTs from September to December (Figure 7.5). This period of high negative correlation lags by approximately 1 month the high correlation found between the JASO mean Nile flow and monthly ENSO index (blue line of Figure 7.4). This lag is consistent with the lag estimated in the cross-correlation analysis between the monthly Nile flow and the monthly SSTs in IO where the peak of the correlation occurs at a lag of approximately 2–3 months while with a similar lag is estimated with the ENSO index of about 2 months as shown in Figure 7.6a. The lagged anti-correlation between the Nile flow and the IO SSTs compared to ENSO is explained by the fact that North and Mid IO SSTs respond to ENSO [Figure 7.6b; Nicholson 1997; Cole et al. 2000]. This lag is consistent with the time required for the warm water to travel from the Pacific to the IO through the ''Indonesian through flow'' and advection by the Indian Equatorial Current [Tomczak and Godfrey 1995].

The cross-correlation in Figure 7.6a should be interpreted with caution. The cross-correlation analysis needs continuous time series, while most of the annual flow in the Nile occurs during the rainy season from June to October. It is unavoidable that the use of the whole monthly flow time series (including the winter season) suppresses the magnitude of the cross-correlation, and results in an apparently lower cross-correlation coefficient. However, these results can still be used to support qualitative reasoning. The peak of the cross-correlation between Nile flow and IO SSTs occurs 1 month later than the corresponding peak between the Nile flow and ENSO. This result indicates that the North and Middle IO SSTs are strongly linked to ENSO, which reveals that any indices based on the North and Middle IO SST cannot be used independent from the ENSO

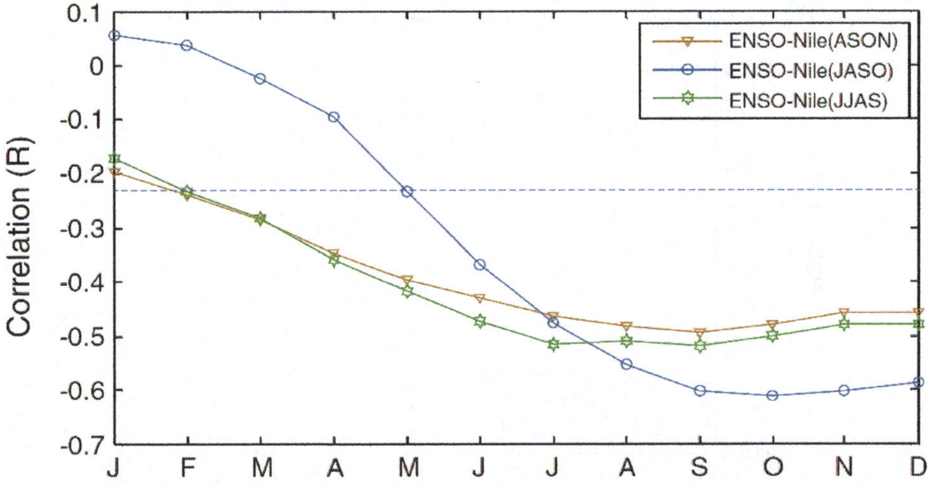

Figure 7.4: Correlation between the monthly ENSO index and the average Nile flows during July–October (JASO), June–September (JJAS) and August–November (ASON) for 1871–2000. The dashed line represents the 1 % significance level of the correlation

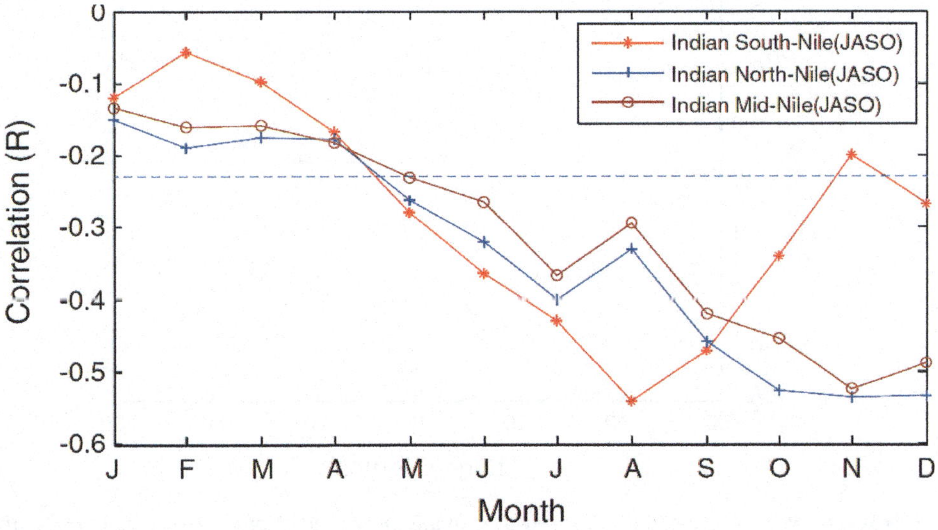

Figure 7.5: (below): Correlation between the monthly IO SSTs and the average Nile flows during July–October (JASO) for 1871–2000. The dashed line represents the 1 % significance level of the correlation

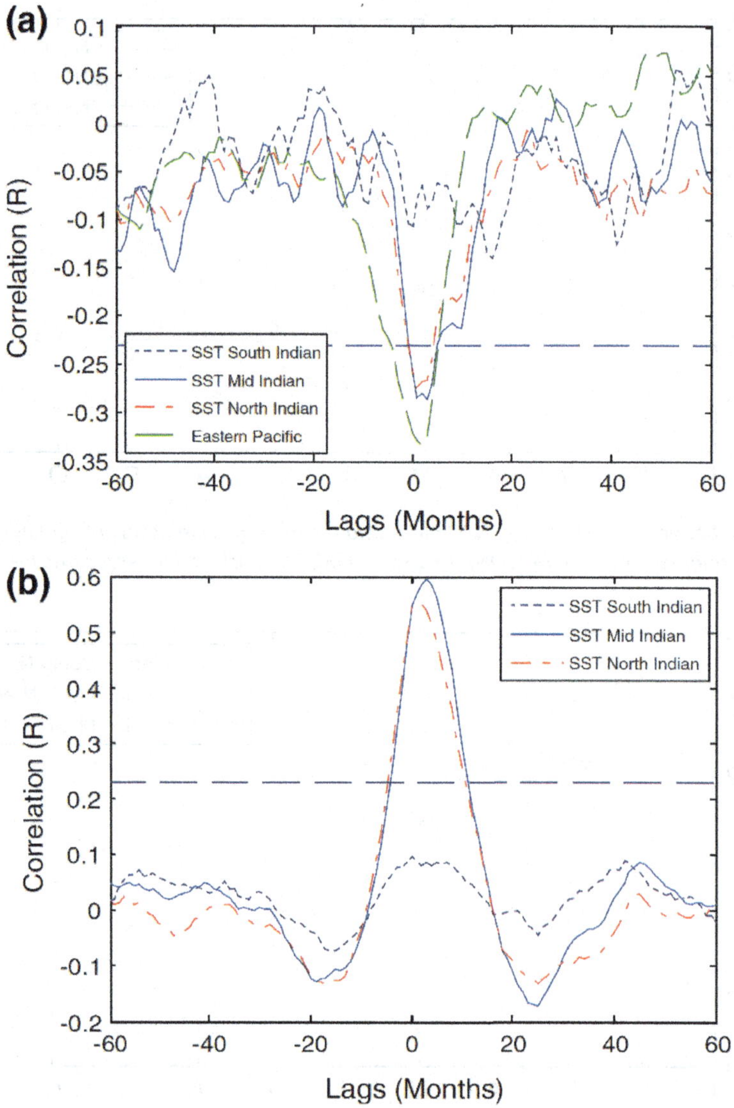

Figure 7.6: (a) Cross-correlation between the monthly SSTs in the three regions of the IO and one region in Eastern Pacific Ocean with the monthly Nile flows. (b) Cross-correlation between the monthly SSTs in the three different regions of the IO and the monthly SST in the Eastern Pacific Ocean. The dashed straight lines represent the 1 % significance level of the correlation

index to predict the Nile flow. On the other hand, using the whole time series, the SST in the South IO shows a low cross-correlation with the Nile flow because significant correlation only occurs during 1 month in August (Figure 7.6a). However, South IO SSTs are significantly less dependent on ENSO. The South Indian Current cools the South IO surface and decreases the effect of the warm South Equatorial Current on the South IO [Tomczak and Godfrey 1995].

Partial coherency analysis

The dependence of the Nile flow on SSTs in the Indian and Pacific Oceans is analyzed in this section using partial coherency analysis. "Partial coherency" represents a generalization of the concept of partial correlation coefficient. Partial correlation analysis focuses on the relationship between two variables in the presence of a third variable [Jobson 1991]. It quantifies the correlation between these two variables after evaluating the impact of the third variable.

In this study, the two variables were the Nile flow and IO SSTs, and the third variable was the ENSO index. The partial correlation coefficient quantifies how much of the Nile flow variability, which cannot be explained by ENSO, is explained by IO SSTs. However, the partial correlation coefficient does not show how this correlation is distributed over different frequencies. We applied the partial coherency analysis, which describes how the partial correlation is distributed among different frequencies. By "partial coherency," we mean the cross-coherency between IO SSTs and the portion of the Nile flow variability that cannot be explained by ENSO.

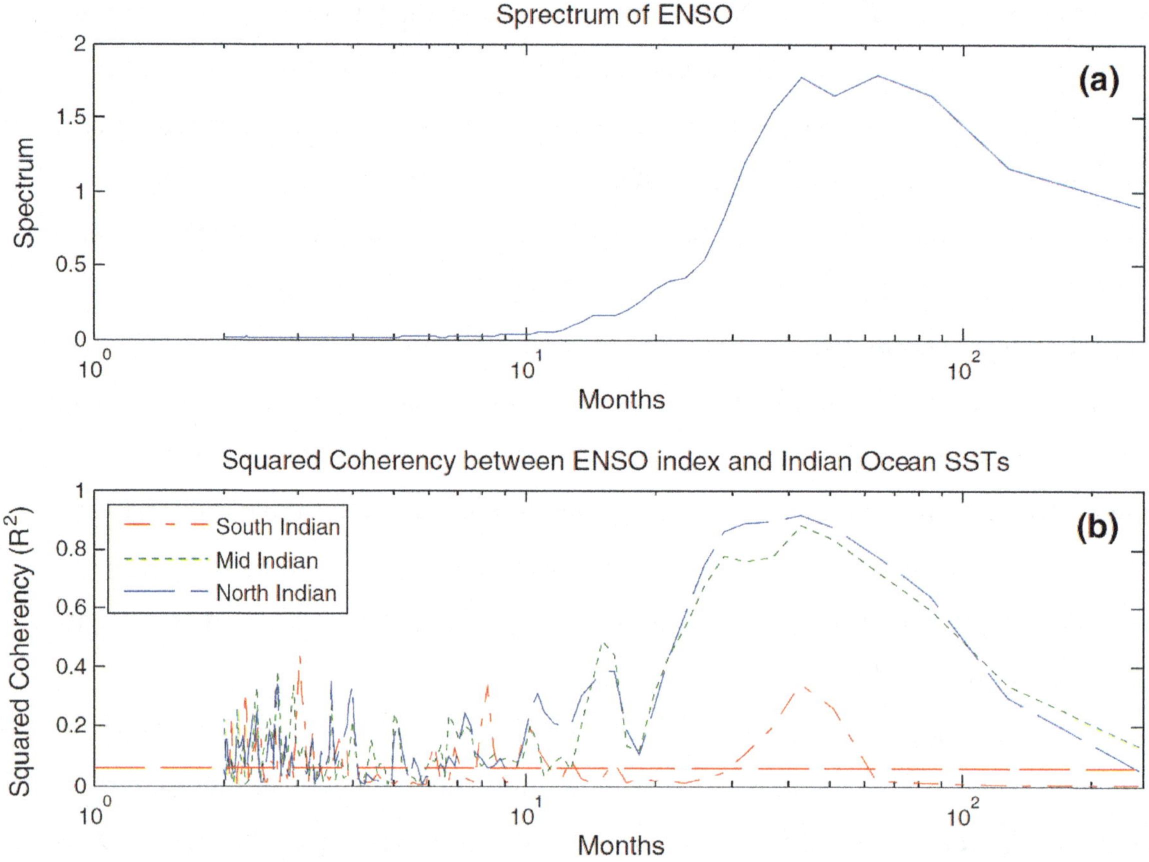

Figure 7.7: (a) Spectrum of ENSO, (b) squared coherency between ENSO and SSTS in the IO. The straight dashed line represents the 1 % significance level of the correlation

The ENSO spectrum, calculated from monthly anomalies, depicts a peak at a time scale of almost 50 months (about 4 years; Figure 7.7a). The squared coherency spectrum, also calculated from monthly anomalies, shows a high coherency between ENSO and North and Middle IO SSTs at the same time scale of 50 months (Figure 7.7b). It shows, however, a low coherency between ENSO and South IO SSTs. This result confirms the cross-correlation analysis illustrated in Figure 7.6a in which South IO SSTs depends less on ENSO than North and Middle IO SSTs.

To verify that IO SSTs cannot explain any additional variability of the Nile flow, we used the partial coherency analysis as described above but this time between the Nile flow and ENSO and IO SSTs. First, we performed 12 linear regressions from January to December between the monthly-mean Nile flow and the monthly-mean ENSO index (u) months earlier, where (u) denotes the possible lag between ENSO and the Nile flow. The linear regressions depended on the time lag (u) according to:

$$R_{new}(i) = a(i) \times E(i-u) + b(i) \qquad (7.1)$$

$R_{new}(i)$is the predicted Nile flow for a specific month (i) where i = 1–12, and $E(i-u)$ is the ENSO index (u) months earlier than month (i). Accordingly, $a(i)$ and $b(i)$ are also functions of (u). The Nile flow residuals that are not explained by ENSO are estimated by:

$$R(i,j) = R_{obs}(i,j) - R_{new}(i,j) \qquad (7.2)$$

$R_{obs}(i,j)$ is the observed Nile flow index in month (i) of year (j), $R_{new}(i,j)$ is the predicted flow in year(j). The new time series describing the flow residual $R(i,j)$ contains the part of the Nile flow variability that cannot be explained by ENSO. The partial coherency is the squared coherency between the time series of the residual flow and the IO SSTs. Because of the dependence of the flow residual on time lag (u) between the Nile flow and ENSO, the partial coherency also depends on (u), but our results indicate that the partial coherency estimates are not sensitive to the change of the time lag

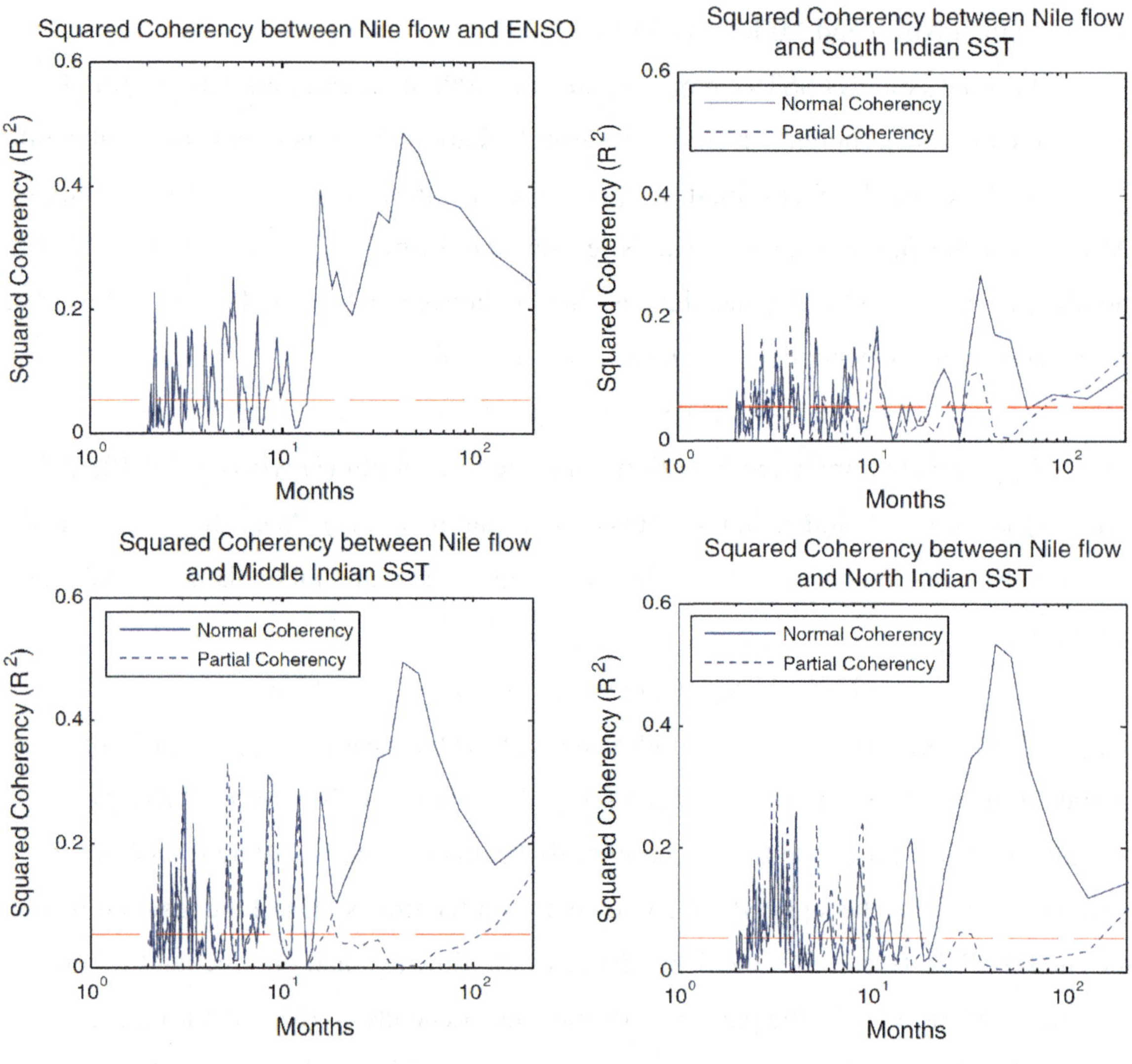

Figure 7.8: Squared Coherency between the Nile flows and SSTs in the Pacific (ENSO index) and IO. The straight dashed lines represent the 1% significance level of the correlation

(not shown). Therefore, the difference between the partial coherences for u = -2, -1, 0, 1, 2 is negligible.

Figure 8 shows the normal and partial coherency spectrum between the Nile flow and IO SSTs. The normal coherency between North and Middle IO SSTs and the Nile flow peaks at the time scale of 50 months (solid lines; bottom panels of Figure 7.8), which is similar to the squared coherency between ENSO and the Nile flow (top left panel). However, the partial coherency is almost zero at the position of this peak (dashed lines). The normal and partial coherencies between the Nile flow and South IO SSTs are both smaller than the normal coherencies between the Nile flow and North and Middle IO SSTs because the correlation has only one peak in August (Figure 7.5). A detailed analysis between the time series of the JASO mean Nile flow and South IO SSTs in August reveals that they are correlated at time scales of 2–4 years (not shown). However, the correlation at the time scale of 4 years is dependent on ENSO, while that at 2 years is independent from ENSO.

These results are consistent with two conclusions:

1. At the 4-year time scale, North and Middle IO SSTs cannot explain variability in the Nile River flow that is not already explained by ENSO. The peak of the normal coherency at the time scale of 4 years can indeed only be explained by a response of North and Middle IO SSTs to ENSO that leads to a response in the Nile flow.
2. The South IO SSTs explain some of the variability of the Nile River flow that is not explained by ENSO. Hence, the South IO is the only region of the IO that

plays a role in shaping the variability of the Nile flow that is independent from ENSO.

Mechanism relating the Indian Ocean SSTs and UBN flow

It is shown earlier in this chapter that the Nile River flow is moderately correlated with the SSTs in the IO. Furthermore, we found that the SSTs in the North and Middle IO respond to the warming in the Pacific, while the SSTs of South IO are less dependent on the Pacific SSTs. Accordingly, we proposed a mechanism that explains these

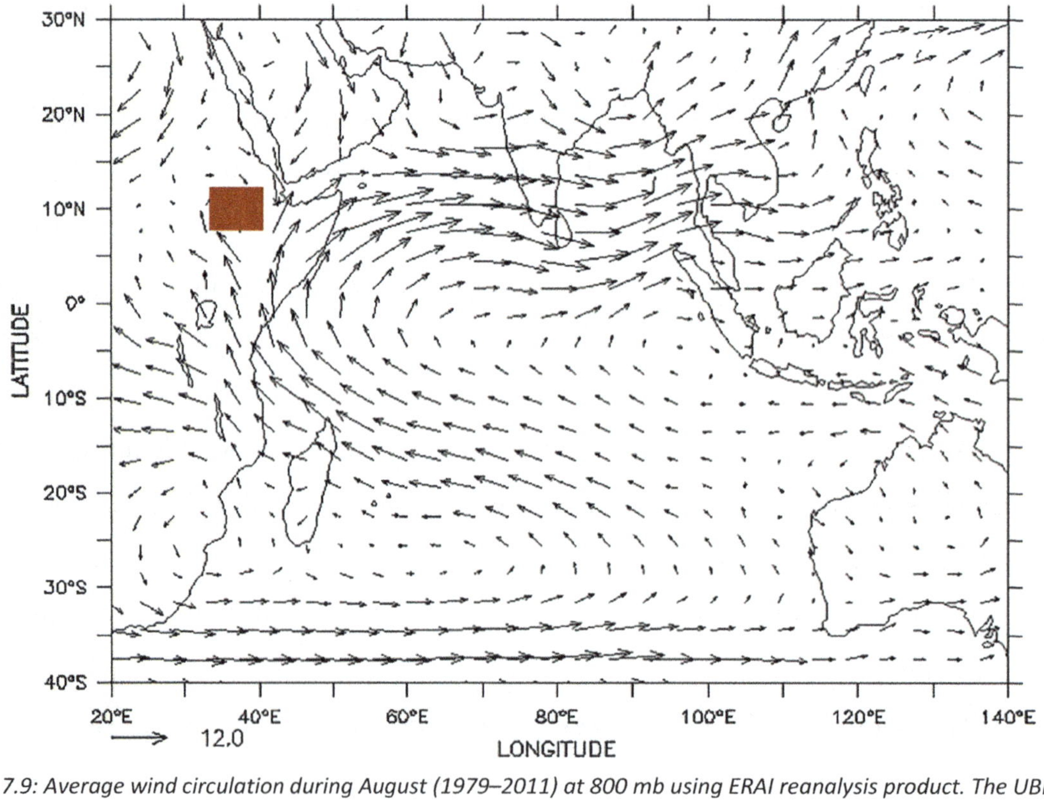

Figure 7.9: Average wind circulation during August (1979–2011) at 800 mb using ERAI reanalysis product. The UBN basin (shaded in brown)

teleconnections between ENSO, IO SSTs and Nile flow. Figure 9 shows the ERAI averaged wind fluxes at 800 mb during August 1979–2011. The wind circulation is mainly composed of two anticyclones in the North and South of the Equator. The anticyclone located in the South is responsible for maintaining the cross-equatorial meridional southerly flow and the associated convergence of air over the UBN basin. The anticyclone located in the North is associated with low level westerly winds that forces divergence of air away from the UBN.

The intensity of the anticyclonic circulations is reflected in the magnitude of the relative vorticity: strong westerly and cross-equatorial southerly low-level flows correspond to high values of relative vorticity over the North and South IO, respectively. It is important to notice that the sign of vorticity in the South of IO is positive (anti-clockwise) and negative (clockwise) in the North of IO (Figure 7.9). The correlation

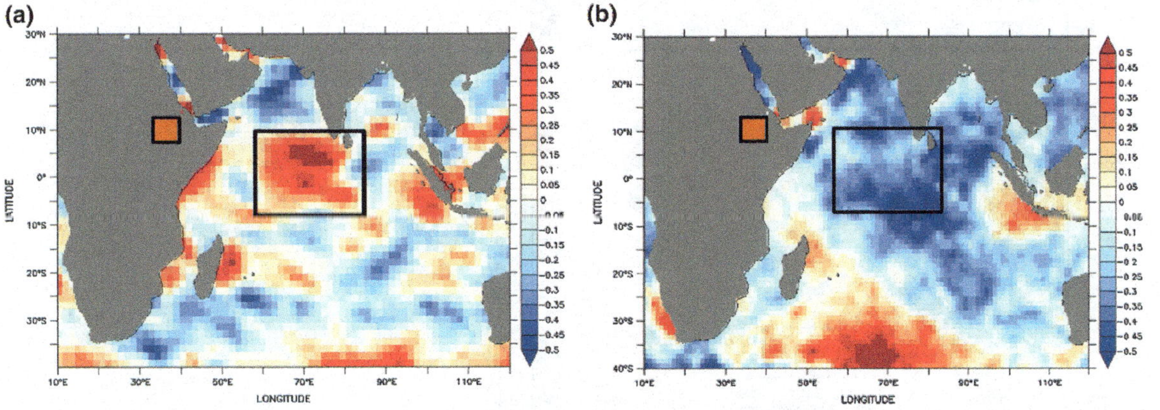

Figure 7.10: a Correlation between the relative vorticity at 700 mb over the IO and averaged convergence of air in the lowest 300 mb over the UBN basin (shaded in brown) during July, and b correlation between average relative vorticity over (60—E–80—E and 5—

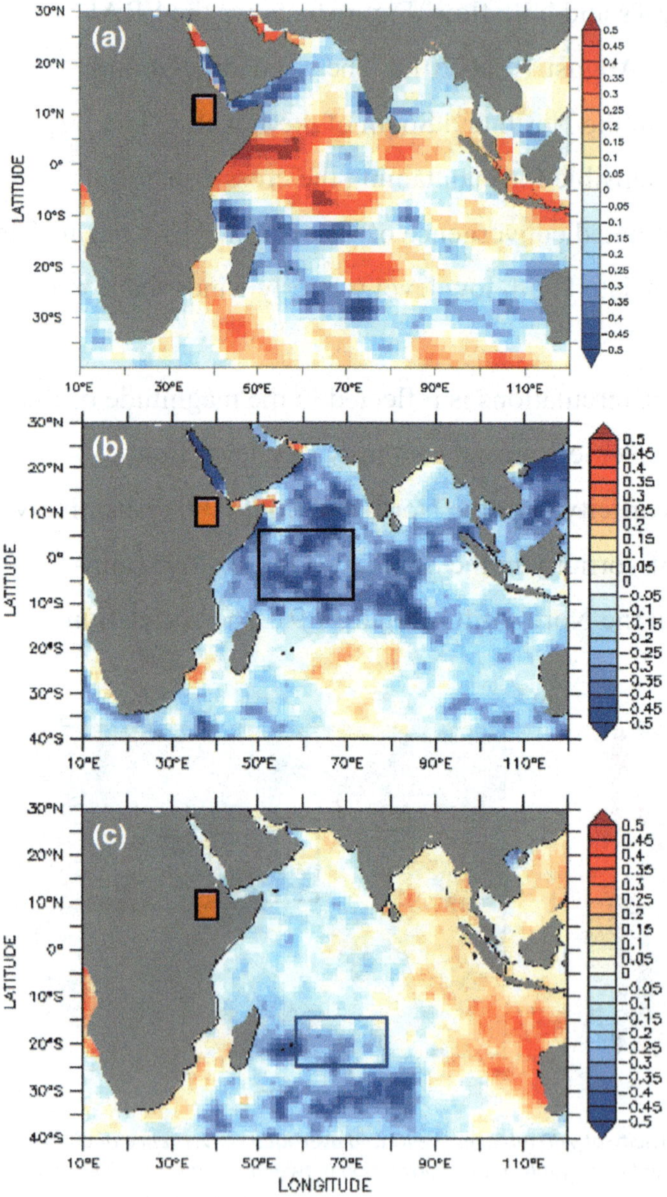

Figure 7.11: (a) Correlation between the relative vorticity at 700 mb over the IO and averaged convergence of air in the lowest 300 mb over the UBN basin (shaded in brown) during August, (b) correlation between average relative vorticity over (50— E–70— E and 5— N–10— S) and SSTs in the IO during August, and c correlation between average relative vorticity over (60— E–80—

between relative vorticity at 700mb and convergence of air in the lowest 300 mb over the UBN basin in July is shown on Figure 7.10a, and that for August is shown on Figure 7.11a. The highest correlation exists over the North IO; the positive sign of the correlation indicates that an increase in convergence of air over the UBN is associated with weaker vorticity over North IO (i.e. the westerly winds have a smaller magnitude). In addition, a significant positive correlation exist in the South (60°E–80°E, 15°S–25°S) in August, which indicates that the convergence of air over UBN increases when vorticity over this region increases (i.e. cross-equatorial meridional southerly from South IO have a higher magnitude).

The magnitude of the relative vorticity over North and South IO is negatively correlated with SSTs in both July (Figure 7.10b) and August (Figure 7.11b). Thus, warming over the IO weakens the anti-cyclonic circulation over the South IO. On the other hand, warming over the North and Middle IO enhances the anticyclonic circulation by inducing westerly winds over the North IO. The anomalous westerlies are induced following a Matsuno–Gill circulation forced by an increase in SSTs around the equator, particularly during El Niño years. The Matsuno–Gill circulation pattern responds to an increase in the IO SSTs in July and August [Yang et al. 2007]. In addition, the increase of the SSTs over the South of the IO, which is independent from ENSO, can similarly generate a cyclonic circulation that enhances the westerly winds over the South IO and reduces southerly cross equatorial flow and the corresponding convergence of air towards the UBN basin.

In Figure 7.12, composite analysis is used to show the difference in winds circulation over the IO during warm and cold SSTs conditions for both the NIO and SIO. In this analysis, the HadISST data is used to identify the warm and cold SSTs conditions and ERAI reanalysis product to provide the corresponding low-level wind circulation. The analysis is limited to the period (1979–2013) as ERAI is only available during this period. Figure 12a, b shows the difference in circulation between the four warmest minus the four coldest SSTs over the NIO and SIO respectively during August. Four events are chosen because they almost represent 10 % of the available events (35 events), which makes each analysis include almost 20 % of the available data (i.e. 10 % warmest and 10% coldest events). Figure 12a shows the increased westerly winds over the NIO due to the warming of the same region, which reduces the convergence of air of the UBN basin and hence the rainfall. On the other hand, Figure 7.12b shows the enhanced cyclonic circulation over the SIO due to warming over that region, which reduces the cross equatorial southerly flow and the associated transport of air and moisture towards the UBN basin and hence reduces the rainfall. These results are consistent with the previous analysis focusing on the relative vorticity and the results in the next section using climate model simulations.

Sensitivity experiments on the response of the UBN flow to Indian Ocean SSTs

In order to analyze the mechanism suggested above as well as the connection between the UBN flow and the IO SSTs, several sensitivity simulations were performed using the UK Met Office Hadley Centre General Environmental Model version 1 (HadGEM1). Complete description of the model can be found in Johns et al. [2006], Martin et al.

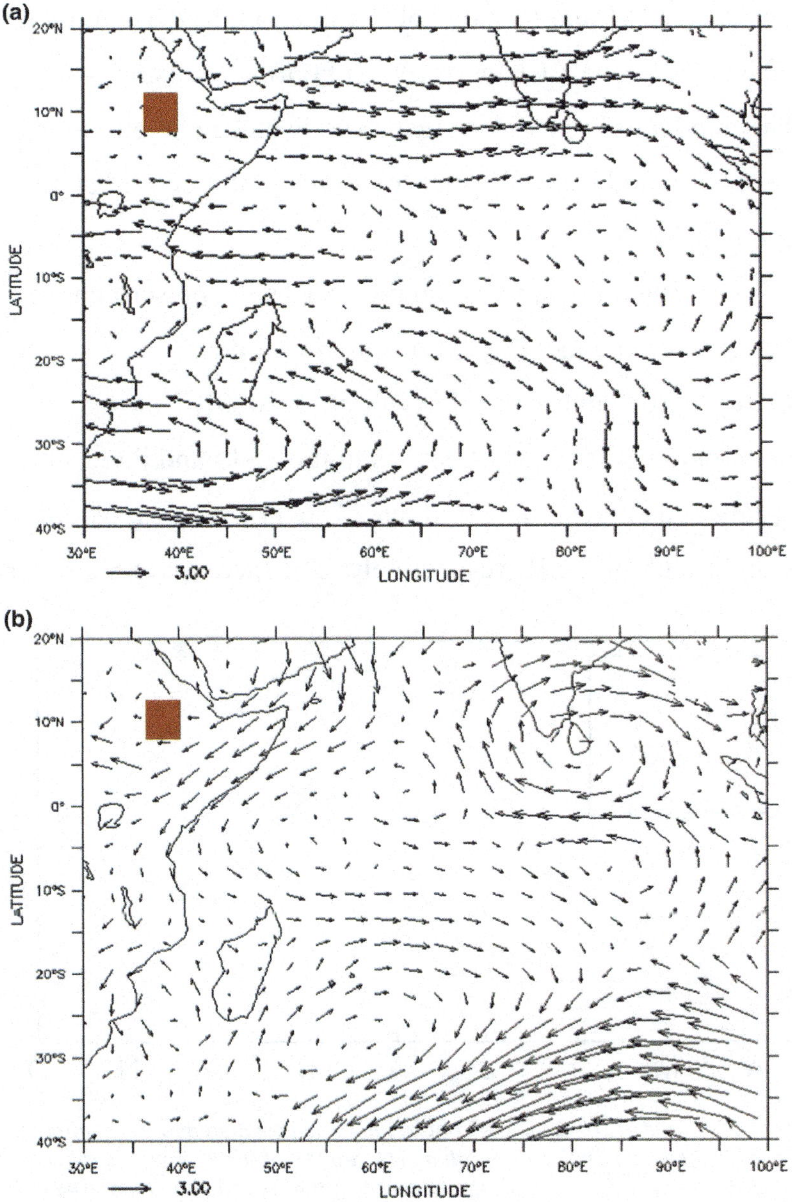

Figure 7.12: Average winds circulation using ERAI reanalysis for the lowest 300 mb from the surface during August for: (a) warm years minus cold years over the NIO, and (b) warm years minus cold years over the SIO. The UBN basin is shaded in brown

[2006] and Ringer et al. [2006]. HadGEM1 and its atmospheric version, HadGAM, took part in the Intergovernmental Panel on Climate Change (IPCC) Fourth Assessment Report (AR4) [IPCC 2007]. This model uses hybrid vertical coordinates in height with 38 vertical levels extending to over 39 km and uses a regular longitude– latitude grid at N96 resolution (approximately 135 km at 50° N).

We performed four different simulations, in which the SSTs were prescribed for 24 years for the period 1979–2002 using the monthly Atmospheric Model Intercomparison Project II (AMIP-II) SST and sea ice [Taylor et al. 2000]. The simulations include a control simulation without any changes to the SSTs and three experiments with SSTs during July and August increased by 2°C over the North/Middle, South and both North/ Middle and South of the IO respectively. The increased SSTs over

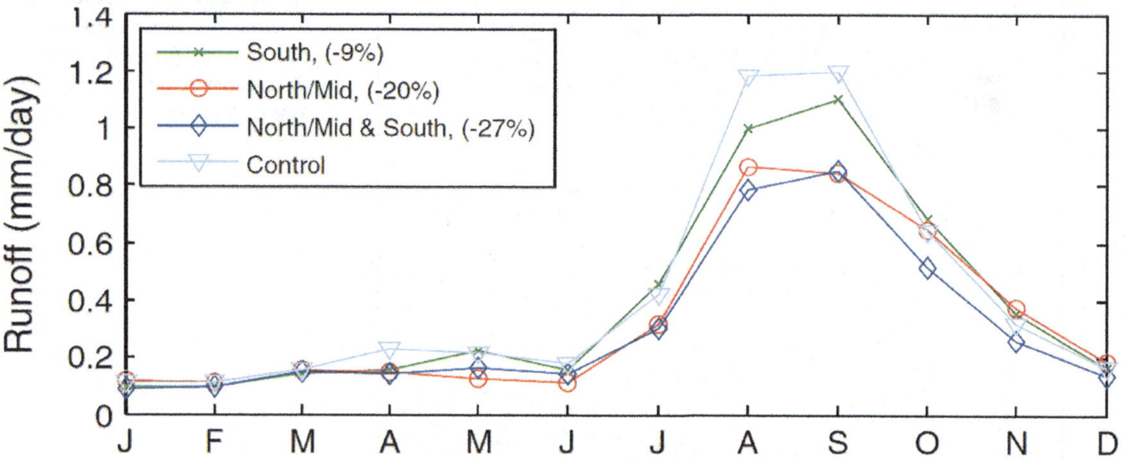

Figure 7.13: Average monthly runoff of the UBN basin (1979–2002) for control simulation compared to three different simulations where the SSTs are increased by 2° C over South, North, South and North/Mid of the IO. The values between brackets are the percentage reduction relative to the mean annual runoff of the control simulation (0.48 mm/day)

the North IO mimic, in distribution, the warming that occurs during El Niño events. It is, however, important to note that 2°C is almost four times the standard deviation of the SST variation in these regions.

The intention for applying such a large SST warming is to force the model to respond strongly, which makes it easier to analyze the results than with a weak warming. Furthermore, as shown in the previous sections, the SSTs are increased over the South IO to show their independent impact from ENSO on the Nile River flow. The monthly mean runoff of the different simulations is shown in Figure 7.13. The model was able to simulate the hydrological cycle of the UBN basin by capturing the seasonal cycle of precipitation (not shown) and runoff (Figure 7.13). In the control simulation, the magnitude of the annual mean precipitation is 3.53 mm/day, which is close to observations (3.3 mm/ day) estimated by the CRU TS3 precipitation dataset [Mitchell and Jones 2005]. Also, the simulated annual mean runoff for the control simulation (0.48 mm/day) represents ~65 % of the observed value for the UBN basin (0.76 mm/ day). The highest decrease in the annual runoff occurs when the SSTs are increased over both the South and North/ Middle of the IO (blue line). However, most of the reduction is related to the increased SSTs over the North/ Middle of IO (red line). The highest decrease of the runoff is relatively large (i.e. 27 %), relative the coefficient of variation in the observed flow ~22 %, which indicates that the UBN flow is sensitive to the warming over selected regions and that SSTs are negatively correlated with the UBN flow. In addition, it is important to notice that we made these experiments with one model; it is possible that other models may have stronger response on the Nile flow for the warming of the IO.

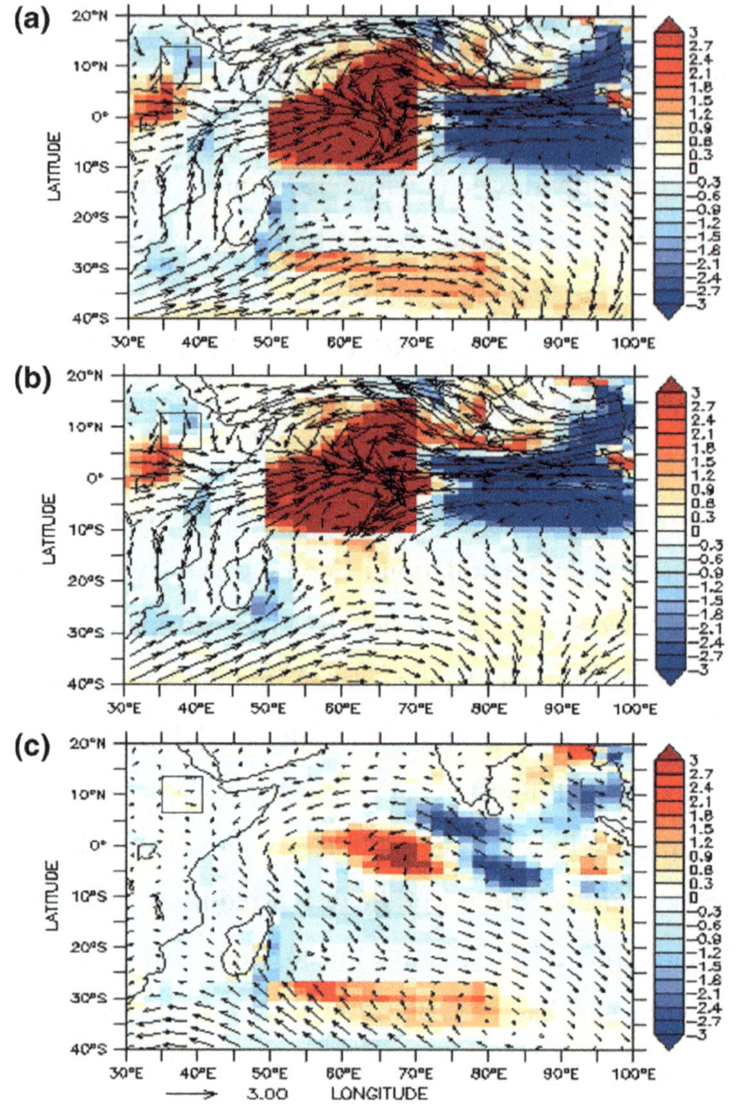

Figure 7.14: Average wind circulation of August for the lowest 300 mb from the surface and precipitation (1979–2002) for perturbed simulation, where SSTs are increased by 2 ——— C over: (a) the North/Mid and South of the IO, minus the control simulation, (b) North/Mid of the IO, minus the control simulation, and (c) South of the IO, minus the control simulation. The small boxes represent the UBN basin.

The corresponding anomalous wind circulations for the simulation in which the SSTs are increased over the North/ Mid and South IO are shown in Figure 7.14. The anomalous westerlies are seen for August along the coast of Africa and over the North/Middle IO following the Matsuno–Gill circulations. These anomalous circulations reduce convergence of air in the boundary layer over the UBN basin and thus the precipitation and the runoff over the basin. The response of the circulation to the warming in the South IO is relatively weak, and fails to reduce the expected impact on the cross-equatorial southerly low-level flow towards the UBN basin. As a result, the magnitude of the impact on rainfall and runoff seems low. The mechanism of the teleconnection between the Nile flow and the ENSO and IO SSTs is summarized in Figure 7.15.

Conclusions

In this study, we investigated the connections between the Nile flow, IO SSTs and ENSO. We showed that SSTs over North and Middle IO respond to the warming of the Pacific Ocean, particularly during El Niño events. Moreover, they do not force any independent significant variability of the Nile River flow. However, the SSTs over the South of IO are less dependent on ENSO and explain some of the variability of the Nile flow that is not explained by ENSO.

We then investigated a mechanism that connects the Nile flow to SSTs in IO and the Pacific Ocean. The increase of the SSTs over the North and Middle of the IO, during El Niño events, forces a Matsuno–Gill circulation over these regions, which enhances westerly winds and reduces the convergence of air over the UBN basin. Similarly, the

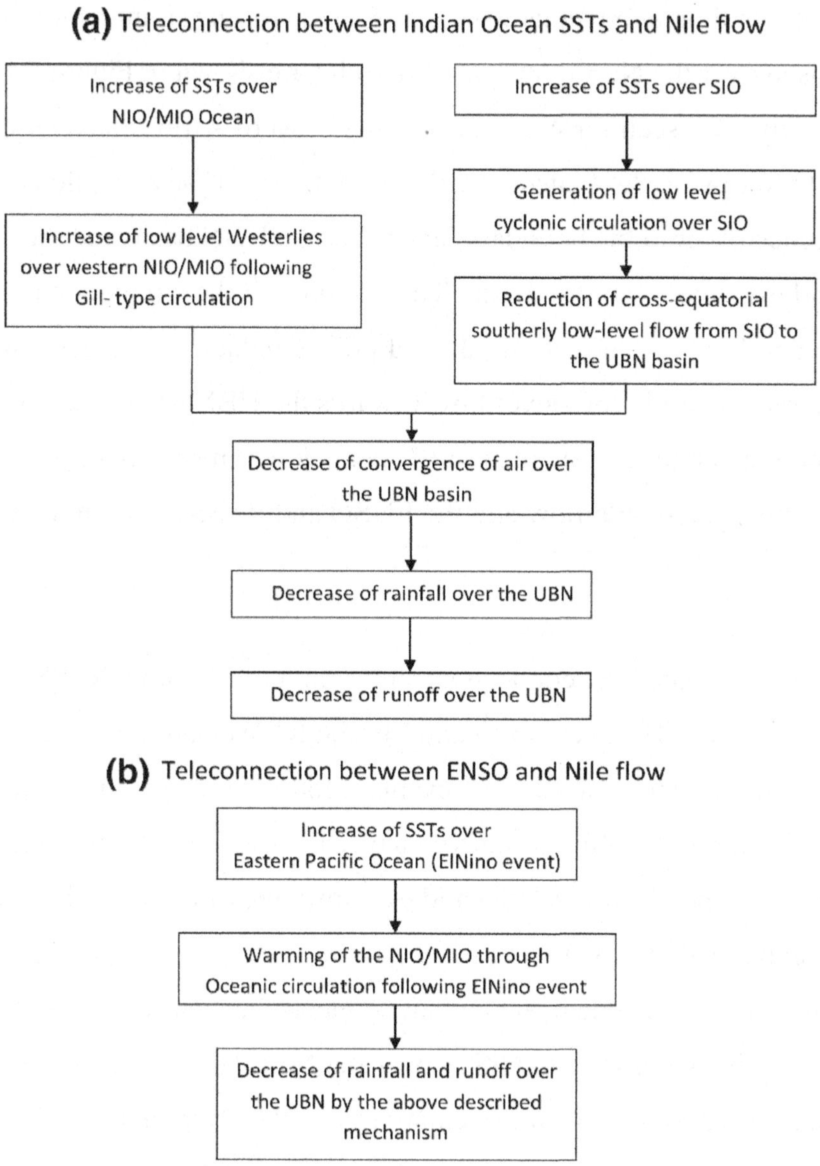

Figure 7.15: Schematic diagram showing the mechanism of the teleconnection

increase in SSTs over the South IO generates an anomalous cyclonic circulation that reduces the cross equatorial southerly low-level flow and the associated convergence of air towards the UBN basin. This reduction in convergence of air is translated to a reduction of rainfall and runoff over the UBN basin. This mechanism was simulated using an atmospheric general circulation model (AGCM) forced with observed SSTs that were modified to simulate a warming over the IO. It is shown that increasing SSTs over IO reduces the Nile flow, which highlights some of the potential mechanisms shaping the impact of climate change on the Nile River flow, through warming of the Pacific and IOs.

A version of this chapter first appeared as:

Siam, M., Wang, G., Demory, M. and Eltahir, E., 2014. Role of the Indian Ocean sea surface temperature in shaping the natural variability in the flow of Nile River. Climate Dynamics, 43(3-4), pp.1011-1023.

Chapter 8—Droughts and floods over the upper catchment of the Blue Nile and their connections to the timing of El Niño and La Niña events

Modathir A. H. Zaroug
Elfatih. A. B. Eltahir

The Blue Nile sustains the life of millions of people in Ethiopia, Sudan and Egypt. Rainfall has a great impact on the social and economic life in the region. Scarcity of rainfall leads to drought, while excessive intense rainfall may lead to flooding. For example, during the 1984 drought in Sudan, Khartoum received only 4.7mm of rain between May and October [Eltayeb, 2003]. This led to crop failure and consequently a famine hit Sudan, leading to massive migration of people in search of food and water [Teklu, 1991].

Floods reflect the other extreme in rainfall fluctuations. There are many factors which affect the severity of a flood, such as terrain slope, soil type and the amount of water in the soil. On 4 August 1988, Khartoum received 210mm of rainfall at Khartoum Central during a 24 h period [Sutcliffe et al., 1989]. (The previous highest daily rainfall

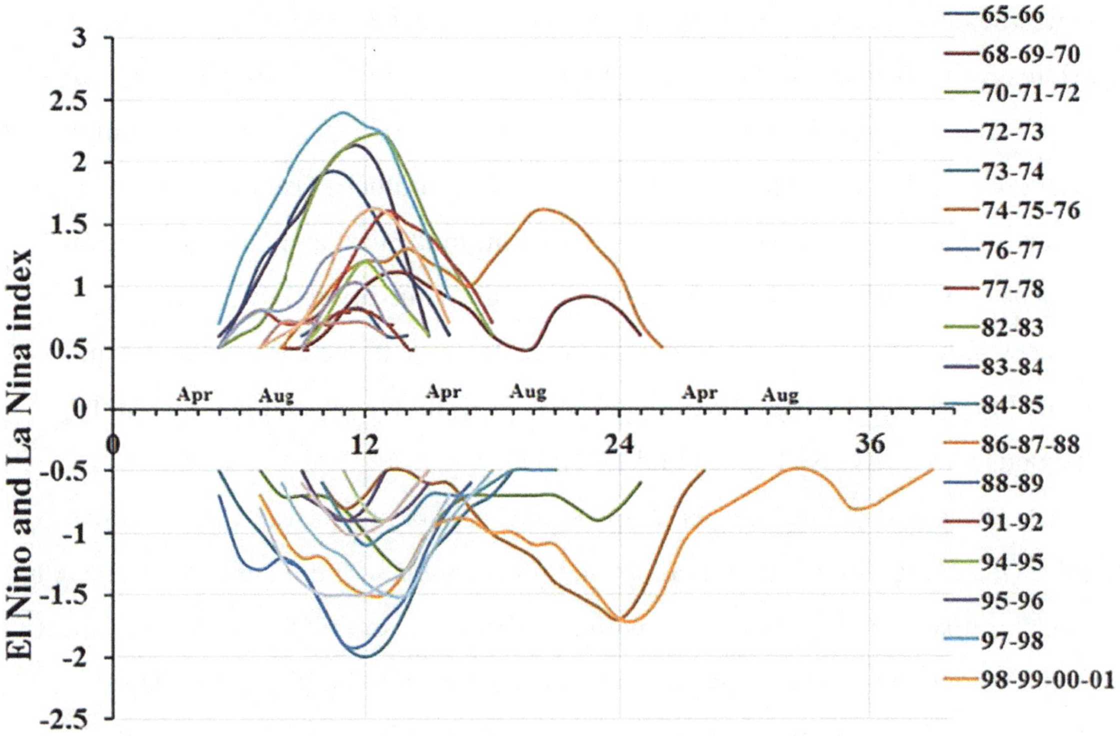

Figure 8.1: El Niño and La Niña timing.

measured at Khartoum since records began in 1899 was 88mm on 31 July 1920. [Hulme and Trilsbach, 1989].) This situation became disastrous when the Nile level also rose about 7m above normal, which led to wide-spread property damage. A rainfall of 200mm at Khartoum was estimated to have a return period of about 500 years by Sutcliffe et al. [1989].

These two natural extreme disasters were associated with significant anomalies in the Pacific sea surface temperature (SST): the El Niño (1983) and La Niña (1988) events. An El Niño/La Niña is the phenomenon in the equatorial Pacific Ocean characterized by a positive/negative sea surface temperature (SST) defined as a 3-month average of SST departures from normal (for the 1971–2000 climatology) in the Niño 3.4 region (5°N–5°S, 120°–170°W) greater than or equal in magnitude to 0.5/−0.5°C for a minimum of 6 consecutive months or longer [Trenberth, 1997]. The Southern Oscillation Index (SOI) is a measure of air pressure difference between Tahiti in the east and Darwin, Australia to the west as compared to the historical average of the same difference. Negative differences indicate El Niño conditions as lower pressure in the eastern Pacific and associated to warmer water and weakened easterly trade winds, and positive SOI corresponds to negative SST index and La Niña.

As discussed through earlier chapters of this book, the last few decades has brought wide recognition that natural oscillations in areas of the Pacific Ocean have a significant impact on the patterns of weather and climate around the world. The most dominant among these oscillations is known as the El Niño–Southern Oscillation (ENSO), which has a return period of about 4 years, varying from 2 to 7 years. Though distant from Africa, ENSO is significantly correlated with rainfall variations over the eastern side of the African continent, but the signs of the correlations and their phase relative to the seasonal cycle vary from region to region [Camberlin et al., 2001].

Aside from evidence presented in this book, Amarasekera et al. [1997] showed that ENSO episodes are negatively correlated with the floods of the Blue Nile and Atbara rivers, which originate in Ethiopia. De Putter et al. [1998] presented a study of decadal periodicities of the Nile River historical discharge of the Roda Nilometer (Cairo, Egypt) and suggested that high-frequency peaks could be linked to ENSO. Abtew et al. [2009]

analyzed monthly rainfall observations from a 32-rain gauge monitoring network in the upper Blue Nile Basin and found that high rainfall is likely to occur during La Niña years and low rainfall conditions during El Niño years. He also found that extremely dry years are highly likely to occur during El Niño years and extremely wet years are highly likely to occur during La Niña years. Finally, Seleshi and Zanke [2004] reported that June–September rainfall in the Ethiopian Highlands is positively correlated to the Southern Oscillation Index (SOI) and negatively correlated to the equatorial eastern Pacific SST. An advantage of knowing the timing of El Niño and La Niña is that they have been shown to be generally predictable for around 6–12 months in advance [Barnett et al., 1988; Latif et al., 1998; Barnston et al., 1994; Chen et al., 2004; Chen and Cane, 2008]. So, our thinking going into the study reported in this chapter was that if we found a strong correlation between ENSO and the flow of the upper catchment of the Blue Nile, this would increase the ability to predict the flow pattern in the Blue Nile.

Several studies attempted to use oceanic and atmospheric variables as predictors in seasonal hydrologic forecasting over East Africa [Mutai et al., 1998; Hastenrath et al., 2004; Philippon et al., 2002; Yeshanew and Jury, 2007; Mwale and Gan, 2005; Williams and Funk, 2011, 2010]; however none of these studies focused on the June–September rainfall in Ethiopia. In this chapter's study, we analyzed river flow and rainfall observations with the goal of evaluating the impact of El Niño on drought and flood conditions in the upper catchment of the Blue Nile. Flood and drought in the context of this paper refer to high and low flows. Not all El Niño and La Niña events are the same (see Figure 8.1)—they have different timings and characters. In fact, different events start

at

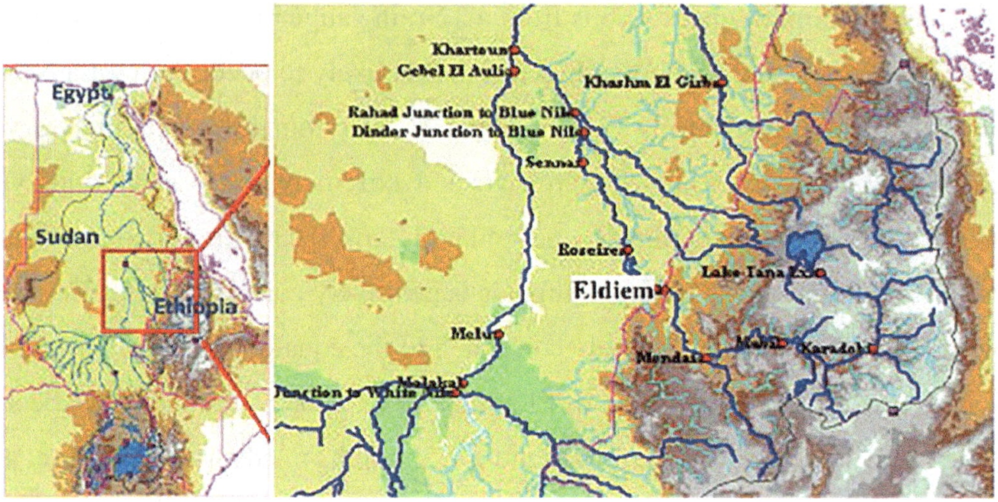

Figure 8.2: The topography and geography of cities in the region.

different times of the year and their sequence exhibits different patterns. In this study, we focused on the dependence of occurrence of droughts and floods in the upper catchment of the Blue Nile on the timing and sequence of El Niño and La Niña events. In particular, we attempted to identify the sequence of Pacific Ocean seasonal SST conditions that significantly affect drought and flood conditions over Ethiopia in order to provide recommendations for possible use as input to seasonal water resources forecasting systems. This would have great economic and social value for the management of water resources in the region.

Data and methods

Discharge measurements between 1965 and 2012 from Eldiem station (Figure 8.2), located at the border between Sudan and Ethiopia about 120 km upstream from Elrosieres dam (Figure 8.2), were used in this study. The gauge station measures water level and discharge at the outlet of the upper catchment of the Blue Nile. The data at Eldiem station from 1997 to 2001 were missing, and these missing data points were filled by using the nearest station to Eldiem, Elrosieres, noting that there are no contributing tributaries between the two stations. The contribution of inflow/losses to the river between Eldiem and Elrosieres is small compared to the contribution of the runoff from rainfall in the whole upper catchment of the Blue Nile. In this study we accumulated the daily discharge of June, July, August, and September at Eldiem station, and the accumulated daily discharge at Elrosieres during the same period is almost similar to Eldiem.

Conway [1997] made a water balance model of the upper Blue Nile in Ethiopia. He used Elrosieres record for 1912–1963, before construction of the Elrosieres dam, and Eldiem for the period 1964–1987 to provide a continuous time series. He found good agreement between two gauges during their period of overlap (annual correlation = 0.97). The discharge data represents the catchment hydrology better than the rainfall data from scattered set of sparsely distributed stations. In fact, Duethmann et al. [2013] concluded that the rainfall data has a relatively large uncertainty due to errors in measurement, wind, and high spatial variability of precipitation in the mountainous regions. The density of rain gauge networks is often low, and the gauges are often unevenly distributed.

For this reason, we used multiple precipitation data sets [Zaroug et al., 2013]: the global data set of monthly precipitation from the Global Precipitation Climatology Project (GPCP) version 2.2 [Huffman et al., 2011], which is a satellite/gauge–merged rainfall product available from January 1979 to December 2010 with a resolution of 2.5; the Climate Research Unit (CRU, land only) $0.5° \times 0.5°$ resolution monthly precipitation data set [Mitchell et al., 2004], which is a purely gridded gauge product; and the University of Delaware (UDEL) monthly global gridded high-resolution station (land) data (0.5 ×0.5 resolution) available from 1900 to 2010 (ESRL, 1900–2010). While it is true that GPCP spatial resolution is relatively coarse, this data set extends for a long period of about 30 years. This is much longer than other data sets such as TRMM, which has a higher spatial resolution but significantly shorter length of about 15 years.

In order to identify El Niño conditions, the Niño 3.4 index between 1965 and 2012 was downloaded from the NOAA website [NOAA, 2014]. The data from the Niño 3.4 region was preliminarily analyzed in relation to several precipitation observational data sets.

Results and discussion

Relation of Pacific SST and discharge at Eldiem Station

Figure 3c shows the discharge at Eldiem station and its association with El Niño and La Niña years, with the upper panels presenting two specific episodes (Figure 8.3a and b). As can be seen from these figures, El Niño years such as 1972 and 1987 are associated with low discharge, while La Niña years, for example 1988, are associated with relatively high discharge. This result is more clearly and quantitatively depicted in Figure 8.4, which shows the monthly discharge (January– December) averaged over all El Niño (1965, 1968, 1969, 1972, 1982, 1983, 1986, 1987, 1991, 1992, 1995, 1997, 2002, 2004, and 2009), La Niña (1970, 1971, 1973, 1974, 1975, 1985, 1988, 1989, 1998, 1999, 2000, 2007, 2008, and 2010) and normal years (1966, 1967, 1976, 1977, 1978, 1979, 1980, 1981, 1984, 1990, 1993, 1994, 1996, 2001, 2003, 2005, 2006, 2011, and 2012). This

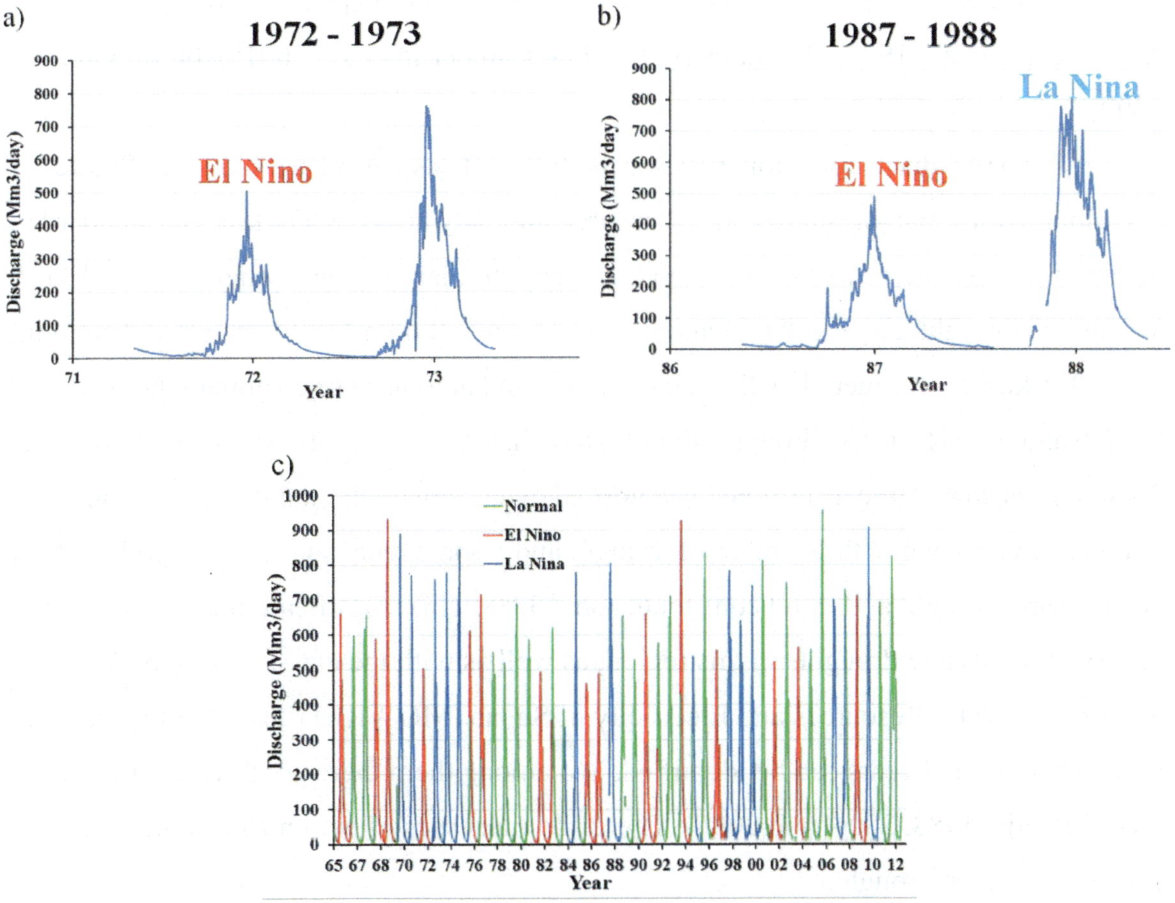

Figure 8.3: The discharge of the Blue Nile at Eldiem station (1965– 2012) and its association with El Niño and La Niña years in (c), red represents El Niño event periods, blue represents La Niña event periods, and green represents normal event periods. (a) and (b) are a zoom on some El Niño and La Niña years.

confirms the results of previous studies that El Niño is mostly associated with below-average rainfall, and La Niña with above-average rainfall [Chapters 2 and 4; Amarasekera et al., 1997; De Putter et al., 1998; Camberlin et al., 2001; Abtew et al., 2009].

The JJAS discharge anomalies for the full analysis period are shown in Figure 8.5, along with some thresholds: any discharge anomaly above 6.971 km^3, 1× standard deviation, is considered as extreme flood, and any discharge anomaly below −6.971 km^3 as extreme drought; any discharge anomaly between 3.486 km^3 (0.5× standard deviation) and 6.971 km^3 is considered as flood conditions, and any discharge anomaly between −3.486 and −6.971 km^3 as drought. Finally, any discharge anomaly between −3.486 and 3.486 km^3 is considered as normal. The normal events cover about 38% of the total number of years, while the number of drought and flood events contributes another 30 %, and extreme drought and flood contribute some 32%. This classification is in line with observed floods and droughts in this region, as well as with the classification of the Ministry of Water Resources and Electricity of Sudan (MOWRE) [2011]. In Figure 8.5, nine extreme flood years can be identified, and among them there are three at or close to record floods (1988, 2006, and 2007). There are nine cases of extreme drought, five floods, and seven droughts.

The coefficient of correlation and coefficient of determination are calculated for the linear fit between Niño 3.4 and discharge anomalies in each season as illustrated in Table 8.1. Figure 6 shows examples of the relation between SST anomalies in the Niño 3.4 region in different seasons and JJAS discharge anomalies at Eldiem station. The

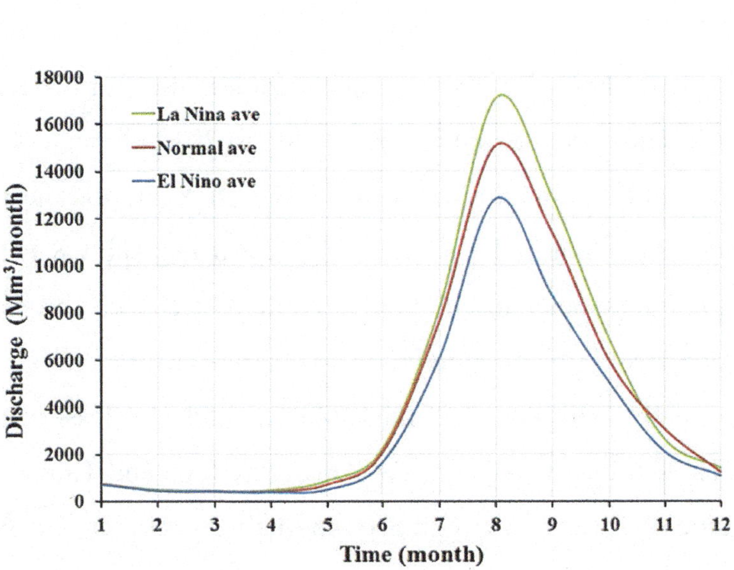

Figure 8.4: Monthly discharge at Eldiem station averaged during El Niño (1965, 1968, 1969, 1972, 1982, 1983, 1986, 1987, 1991, 1992, 1995, 1997, 2002, 2004 and 2009), La Niña (1970, 1971, 1973, 1974, 1975, 1985, 1988, 1989, 1998, 1999, 2000, 2007, 2008 and 2010) and normal years (1966, 1967, 1976, 1977, 1978, 1979, 1980, 1981, 1984, 1990, 1993, 1994, 1996, 2001, 2003, 2005, 2006, 2011 and 2012).

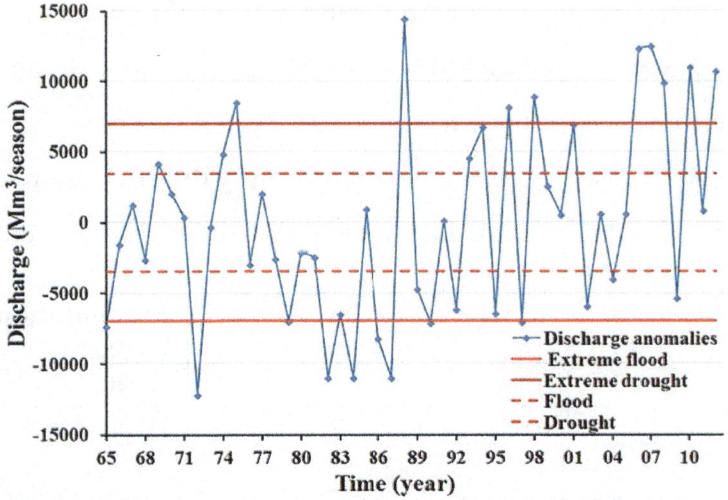

Figure 8.5: The discharge anomalies at Eldiem station averaged over JJAS (1965–2012), the red line represents the threshold for the extreme flood/drought, and the dashed red line represents the threshold for drought/flood.

scatter plots are made for Niño 3.4 versus discharge anomalies for different seasons as shown in Figure 8.7. The period extended from 1965 to 2012. A negative correlation between the Niño 3.4 SST anomalies in (JFM, AMJ, and ASO) and the JJAS discharge anomalies at Eldiem is evident in the panel of Figure 8.6 and Table 8.1. For example, the large El Niño of 1987 is associated with below-average discharge and the La Niña of 1988 is clearly associated with above-average discharge. This negative correlation is less evident in the case of JFM (Figure 8.6a and Table 8.1) and, to a lesser extent, AMJ (Figure 8.6b and Table 8.1), and higher in ASO (Figure 8.6c and Table 8.1) SST anomalies. The same plot was made with SST anomalies for other seasons; FMA, MAM, MJJ, JJA and JAS (not shown here), and FMA and MAM also showed lower correlations compared to the MJJ, JJA and JAS anomalies.

The highest correlation was found during JJA and JAS (−0.56, not shown here). Figure 6 thus illustrates that the rainfall in the upper Blue Nile River catchment is highly sensitive to the SST during AMJ to ASO. In this analysis we look at the correlation and coefficient of determination for the whole period (1965–2012) regardless of El Niño, La Niña or normal years. These results are not significantly different from the results presented in Chapter 2, which detailed the annual flow in the Nile River at Aswan from

SST index	**JJAS R^2**	**Precipitation correlation**
JFM	0.01	–0.08
AMJ	0.15	–0.39
ASO	0.29	–0.53

Table 8.1: The coefficient of determination and correlation for the linear fit between Niño 3.4 index in different seasons and the JJAS discharge anomalies at Eldiem station for the period 1965–2012.

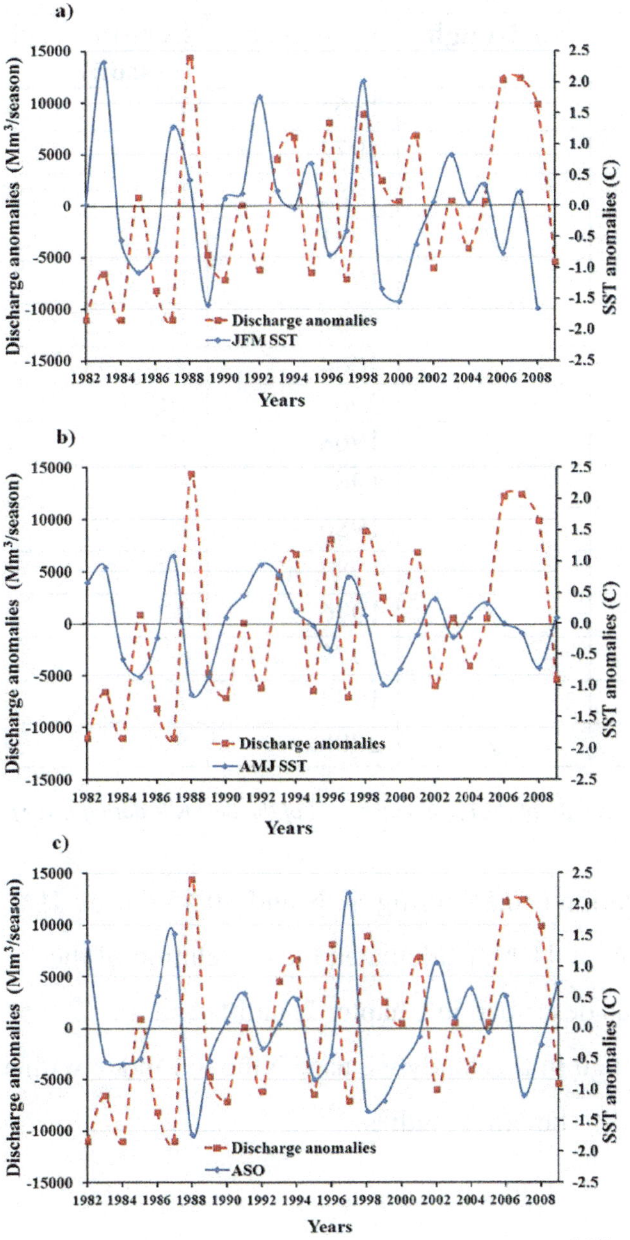

Figure 8.6: The SST anomalies during (a) JFM, (b) AMJ, and (c) ASO in Niño 3.4 region and the discharge anomalies in Eldiem station during JJAS from 1982 to 2009.

Start of El Niño	Extreme drought	Drought	No drought	Flow year	Length (season)
AMJ (1965)	✓			1965	12
AMJ (1972)	✓			1972	11
AMJ (1982)	✓			1982	14
AMJ (1991)			✓	1991	14
AMJ (1997)	✓			1997	12
AMJ (2002)		✓		2002	10
JJA (2004)				2004	7
JJA (2009)				2009	10
JAS (1968)			✓	1968	18
	✓		✓	1969	
JAS (1986)	✓			1986	19
	✓			1987	
ASO (1976)			✓	1976	6
ASO (1977)			✓	1977	6
ASO (1994)			✓	1994	7
ASO (2006)			✓	2006	5

Table 8.2: The effect of the start date of El Niño on the drought of the upper catchment of the Blue Nile during JJAS of the same year.

1872 to 1972 and found the highest correlation (−0.5) during SON and −0.45 during JJA with different SST regions. Amarasekera et al. [1997] calculated the correlation of the Blue Nile using the same SST index as was described in Chapter 2, and found the correlation −0.44 for JJA and SON. Although in this study we used Niño 3.4 index with a different period, our results were in line with the other studies.

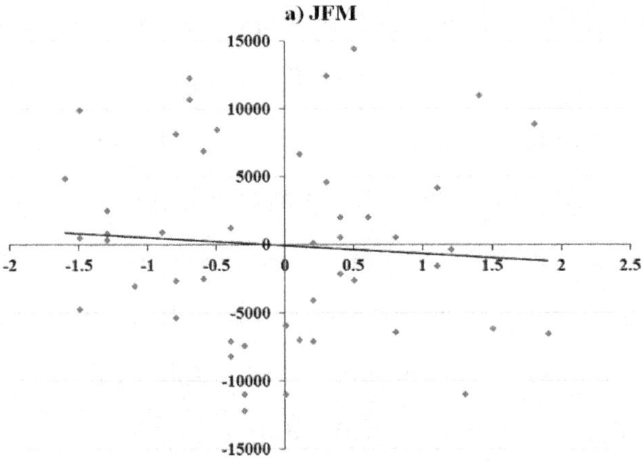

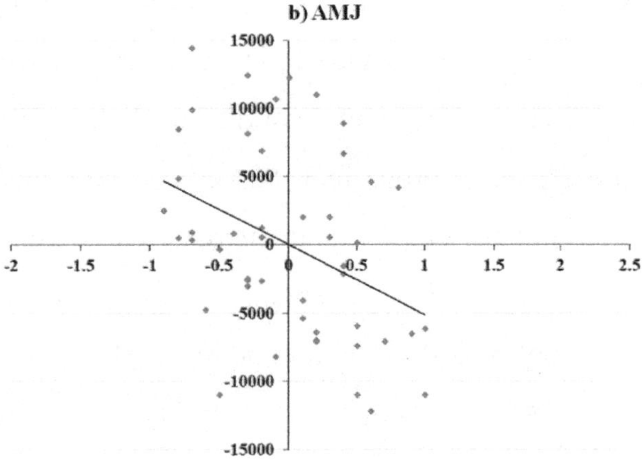

Figure 8.7: The scatter plots for the discharge anomalies at Eldiem station versus Niño 3.4 during (a) JFM, (b) AMJ, and (c) ASO, for the period 1965–2012.

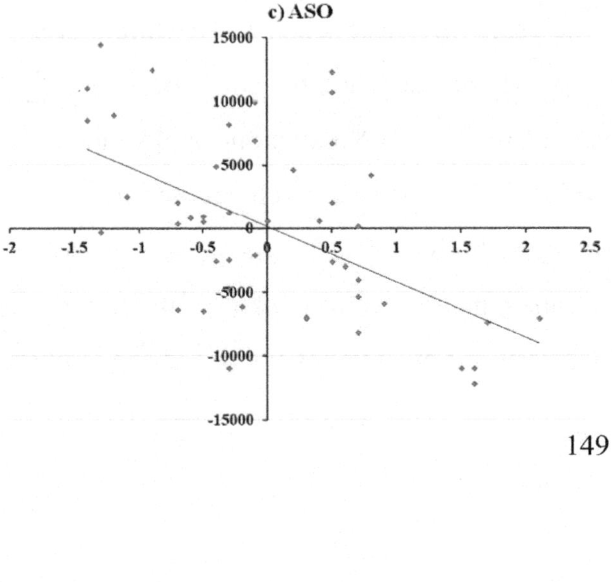

The impact of the start date of El Niño on the drought of the upper catchment of the Blue Nile is further illustrated in Table 8.2. The first column in Table 8.2 shows the starting season of El Niño, the second and third columns then indicate whether there was an extreme drought or drought episode during the same year (JJAS) over the upper catchment of the Blue Nile, while the fourth column shows whether there was no drought. The flow year column shows the start year of each El Niño event, while the length column refers to the duration of the El Niño episode expressed in number of months. Table 8.2 shows that for the six episodes in which El Niño started in AMJ, an extreme drought occurs four times, and one time drought conditions prevails, with only one year having normal conditions. When El Niño starts in AMJ the likelihood of occurrence of droughts is 83%. We acknowledge that this percentage is estimated based on a limited sample of six events only. When El Niño starts in JJA, there are two droughts out of two events. More mixed results are found when El Niño starts in JAS (cases of both drought and no drought equally distributed). Finally, when El Niño starts late in ASO, it tends to be relatively short (1976, 1977, 1994, and 2006), and for the years available there is no drought event (in the same year) for four occurrences (while there was one case of flood and one of extreme flood). However, when the correlation was calculated for the all years of ASO regardless of El Niño, La Niña and normal years during ASO, it was −0.53 as illustrated in Table 8.1. Eltahir's research detailed in Chapter 2 and Amarasekera et al. [1997] also found high correlation during this season. In this study, however, the added value of our analysis pointed to the fact that when El Niño starts in ASO, it has no impact on the Blue Nile flow. The results of Table 8.2 thus

Start of La Niña	Extreme flood	Flood	No flood	Flow year	Length (season)
AMJ (1973)			✓	1973	36
		✓		1974	
	✓			1975	
AMJ (1988)	✓			1988	13
JJA (1970)			✓	1970	18
			✓	1971	
JJA (1998)	✓			1998	33
			✓	1999	
			✓	2000	
JJA (2010)	✓			2010	10
JAS (2007)	✓			2007	11
	✓			2008	
ASO (1983)			✓	1983	5
ASO (1995)			✓	1995	7
ASO (2011)			✓	2011	7

Table 8.3: The effect of the start of La Niña in the flood of the upper catchment of the Blue Nile during JJAS of the same year.

suggest that there is a relation between El Niño events that start in AMJ and drought conditions in Ethiopia, while no effect is found when El Niño events start late in the year in ASO.

La Niña is normally associated with floods in the upper catchment of the Blue Nile [Chapters 2 and 4; Amarasekera et al., 1997]. In Table 8.3 the role of the start date of the La Niña season is explored in terms of its relation with flood conditions in the upper catchment of the Blue Nile (in the same year). The first column in Table 8.3 shows the season of the start of La Niña, and from Table 8.3 it is clear that La Niña events can last for up to three years, as in 1973–1975 and 1998–2000. When La Niña started in AMJ of 1988, there was one extreme flood (in the same year), when it started in AMJ of 1973 and extended for 3 years, there was no flood (in the same year), and one flood and one extreme flood in the following years. When La Niña started in JJA, there was no flood in 1970 and there were extreme flood conditions in 1998 and 2010. When La Niña started in JAS of 2007, there was an extreme flood. When La Niña started late in ASO, it tended to be also relatively short (1983, 1995, and 2011), there were no floods recorded, and in one case (2011) there was even a strong drought. Similar to the analysis of the start of El Niño during ASO, there is no impact on the upper Blue Nile flow. Therefore, in general, when La Niña started in AMJ, JJA and JAS, 67% of the time there was a flood or extreme flood, showing that the rainfall and the monsoon in this catchment is sensitive to AMJ, JJA and JAS SST in the Pacific Ocean. The 67% is calculated as follows:

$$\frac{4(\text{extreme flood or flood events during the first year of La Niña event})}{6(\text{number of La Niña events})} = 67\% \qquad (8.1)$$

In this analysis it is important to highlight the limited number of events that are used to calculate this percentage. In this paper we also explored the importance of the sequence of El Niño followed by La Niña in relation to flood conditions in the upper Blue Nile catchment. For the last 40 years when El Niño was immediately followed by La Niña conditions, there were extreme flood records in the upper catchment of the Blue Nile (1988, 1998, 2007, and 2010), in other words, 67% of the cases of extreme flood

Year	DJF	JFM	FMA	MAM	AMJ	MJJ	JJA	JAS	ASO	Remark
1970	–						+			Normal (above average)
1973			–		+					Normal
1988		–			+					Extreme flood
1998				–			+			Extreme flood
2007	–							+		Extreme flood
2010				–			+			Extreme flood

Minus (–): End of El Niño.
Plus (+): Start of La Niña.

Table 8.4: El Niño followed by La Niña and occurrence of extreme flood conditions.

(Table 8.4). The minus sign in Table 8.4 represents the end of the El Niño period, and the plus sign represents the start of La Niña. If we look at the period from the 1980s to present, it can be concluded that when El Niño is followed by La Niña, in the four recent sequential events there was extreme flood in the Blue Nile. It is an important result with crucial implications for water managers when El Niño quickly followed by La Niña—for example El Niño of 1998 and 2010 ended in MAM and La Niña started quickly in JJA with an extreme flood. It should be emphasized, however, that these results are based on a limited sample size (only six events). In this analysis we excluded the events of 1983 and 1995, because La Niña started late in ASO. From the previous analysis in Tables 8.2 and 8.3, it is evident that when El Niño or La Niña starts late in the year, it does not impact rainfall in the upper Blue Nile catchment.

This study highlighted for the first time the likelihood of the occurrence of record flood flow in the upper catchment of the Blue Nile when El Niño is followed by La Niña. However, we must consider the limitation of the few cases studied. The strength of El Niño seems to have no impact on the extreme flood, because, for example, the 1997–1998 El Niño was one of the highest El Niño, but 2006–2007 El Niño was a normal El Niño.

Other SST anomalies outside of the Pacific Ocean may have an impact in the Blue Nile flow; they may enhance or reduce the precipitation in the upper Blue Nile. The earlier studies which found the connection between the Nile floods and summer monsoon rainfall over India was examined in Bhatt [1989] and Whetton and Rutherfurd [1994]. Camberlin [1997] found a teleconnection between the Indian monsoon and the summer rainfall in East Africa, and the predictors for the Indian monsoon can themselves be used as predictors for East African rainfall. Negative sea level pressure anomalies in Bombay are followed by abundant rainfall in Ethiopian Highlands, while positive sea level pressure anomalies lead to below-average rainfall.

Awadalla and Rousselle [1999] attempted to predict the Nile River inflows to the High Aswan Dam (HAD) in Egypt. They found significant improvement in model skill by incorporating, beside the Pacific and Atlantic SST, the Indian SST. Eldaw et al. [2003] showed that the Indian Ocean's SSTs and the Blue Nile River flows are generally negatively correlated, but sometimes, certain regions of the Indian Ocean (e.g., the Arabian Sea and the sea north of Australia) are positively correlated. Seleshi [1991] found that one of the causes of Ethiopian rainfall is the strong movement of moist air

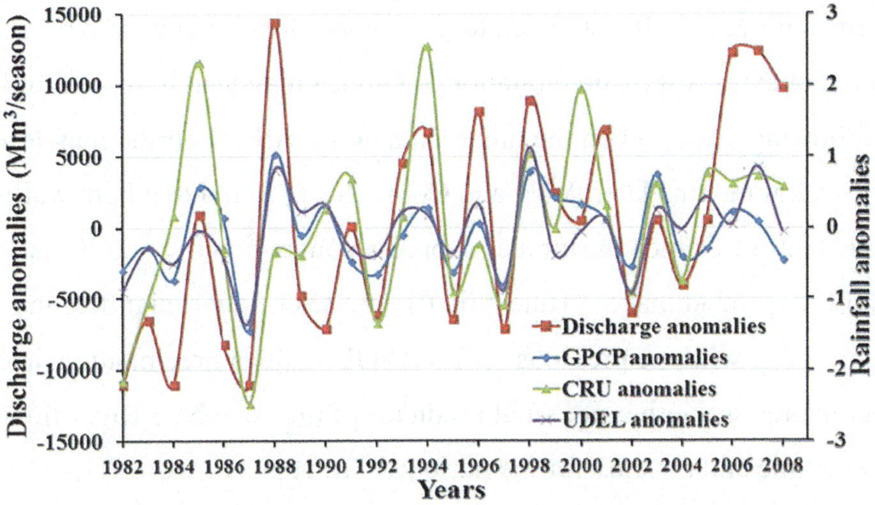

Figure 8.8: Discharge anomalies at Eldiem station and GPCP, CRU and UDEL rainfall anomalies and over Ethiopian Highlands (35° E, 40° E, 8° N, 13° N) during JJAS from 1982 to 2008.

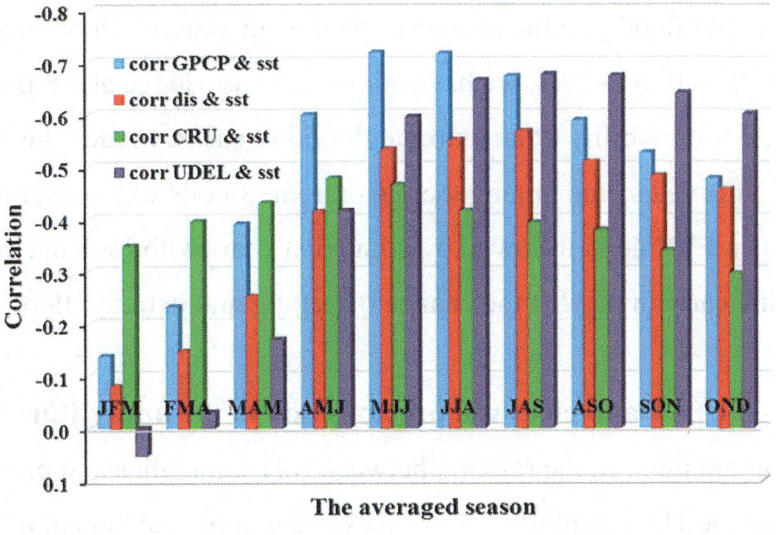

Figure 8.9: Correlation between SST anomalies in Niño 3.4 region and the discharge anomalies at Eldiem station and the GPCP, UDEL and CRU rainfall in the upper catchment of the Blue Nile in the Ethiopian Highlands from 1982 to 2009.

from the high southwest Gulf of Guinea to the low northeast center of Arabia. Gray et al. [1992] found that extra ASON precipitation at Guinea may lead to more rain in the Sahel during the following year, and on the other hand, a dry ASON period may lead to drought in the Sahel several months later. Vizy and Cook [2001] found that both warming and cooling of the Gulf of Guinea in summer suppress convection over northeast Africa. Eldaw et al. [2003] found that the Blue Nile River JASO flow is significantly and positively correlated with the previous year's ASON Guinea precipitation, and the Guinea precipitation is another potential predictor of the Blue Nile River flows, with 11 months of lead time and $r = 0.63$ for the period 1953–1989.

The following example illustrates the added value of knowing the timing of El Niño and La Niña for predicting extreme floods. In the 48 years of analysis (1965–2012) there were 9 extreme floods, so the chance of having an extreme flood in any year during this period was 19%. If, however, we have additional knowledge about the occurrence of a La Niña year, this possibility of an extreme flood increases. In fact, during this period we have 14 La Niña years, and among them 6 extreme floods were observed. As shown in Table 8.4, when El Niño is followed by a La Niña year (with La Niña not starting late in ASO or ending early in MAM) the chance of getting and extreme flood increased to 67%.

Relation of Pacific SST and observed precipitation in the upper Blue Nile catchment

We now turn our attention to the relation between SST anomalies and precipitation. Figure 8.8 shows the JJAS rainfall anomalies over the upper catchment of the Blue Nile from 1982 to 2008 along with the discharge anomalies at Eldiem station. A varying

	GPCP and discharge	CRU and discharge	UDEL and discharge
Correlation	0.64	0.56	0.74

Table 8.5: The correlation between the discharge anomalies at Eldiem station and GPCP, CRU and UDEL rainfall anomalies and over the Ethiopian Highlands (35° E, 40° E, 8° N, 13° N) during JJAS from 1982 to 2008.

correlation between GPCP, CRU, UDEL and discharge anomalies is found for 64, 56, and 74%, respectively (Table 8.5), although the extreme discharge floods in 1988, 2006, 2007 and 2008 appear underestimated in all the rainfall data. This correlation indicates that the GPCP, CRU and UDEL data sets are generally representative of the precipitation variability over the region. Figure 8.9 shows the correlation between GPCP, CRU and UDEL precipitation anomalies over the Ethiopian Highlands and Eldiem discharge with the Niño 3.4 SST anomalies for the entire analysis period and for different seasons. The CRU rainfall anomalies which showed the lowest correlation (Table 8.5) with the discharge anomalies at Eldiem station, it showed the highest correlation with Niño 3.4 index during the early seasons (JFM, FMA and MAM), Eldiem station showed the lowest correlation and insignificant correlation during this period.

However, during MJJ up to ASO, CRU showed the lowest correlation with Niño 3.4, whereas the other rainfall (GPCP and UDEL) and discharge data set showed a higher correlation with Niño 3.4 anomalies during the same period (MJJ–ASO). The correlations between GPCP and UDEL rainfall anomalies and Niño 3.4 index are maximum in magnitude in the AMJ through ASO season compared to the Blue Nile flow at Eldiem station. The correlations are higher for the precipitation than for the discharge except for the CRU data set. There is thus a potential use for the precipitation data set in the hydrological forecasting. Therefore, the ENSO information, with the use of precipitation anomalies, may improve the hydrological prediction.

The corresponding two-tailed t test values are then reported in Figure 8.10, which also gives the threshold for statistical significance at the 95% confidence level. We found a negative correlation in all seasons, indicating that a positive (negative) SST anomaly, or El Niño (La Niña) conditions, tends to lead to drought (flood) conditions. The correlations are maximum in magnitude in the AMJ through ASO seasons, or the late spring late summer period, and tend to decrease in the earlier and later seasons. Also, the correlations are higher for precipitation than for discharge except for the CRU data set during summer, and they show a different seasonal peak (MJJ for GPCP, AMJ for CRU and JJA for UDEL and discharge). Figure 8.10 shows that for all these seasons the correlations are significant at the 95% confidence level, with higher significance for precipitation. These figures thus confirm the strong effect of El Niño anomalies on the hydrology of the Ethiopia Highlands, which feed the Blue Nile River.

Summary and conclusions

Rainfall has a great impact on the social and economic life in the Ethiopian region and upper Nile catchment. Scarcity in rainfall leads to drought while excessive, intense rainfall may lead to flood. Ethiopian rainfall is highly variable, both temporally and spatially, and the rainfall seasonality varies greatly from one region to another [Gissila et al., 2004]. The Blue Nile contributes about 60–69% to the main Nile discharge [Dumont, 1986].

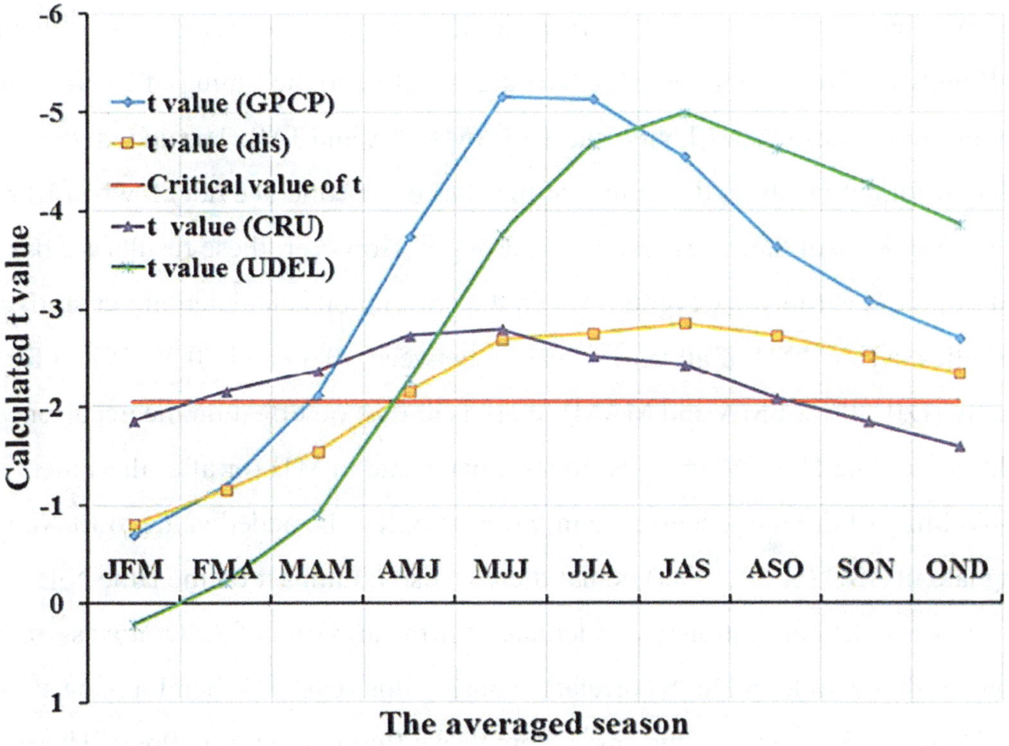

Figure 8.10: 95% significance test of the correlation between SST anomalies in Niño 3.4 region and the GPCP, CRU, UDEL and discharge.

Compared to previous studies, our analysis highlighted the impact of timing and sequence of El Niño and La Niña on the drought and flood conditions over the upper catchment of the Blue Nile. The paper detailed in this chapter also highlighted the role of Pacific SST anomalies in shaping the potential predictability of rainfall over tropical East Africa in both observational discharge at the mouth of the upper catchment of the Blue Nile and different observed precipitation data sets.

We found that that ENSO exerts a significant influence on the upper catchment of the Blue Nile. Droughts in the Blue Nile are sensitive to the timing of El Niño, with 80% of drought cases when El Niño starts in AMJ, JJA and JAS. The added value of including information about the timing of the start of El Niño is evident; when El Niño starts in AMJ, 83% of the cases resulted in drought. However, these results are based on a limited sample size (only six events). When the correlation calculated in our study and other studies during AMJ of all years it varied between −0.36 and −0.39. When El Niño ends early (DJF, JFM, FMA and MAM), there is almost no effect on the occurrence of drought in the Blue Nile. When El Niño terminates late in MJJ (or after that) there is a high possibility of drought occurrence in the Blue Nile. The added value to El Niño starting late in ASO (or after that) is that there is also no impact on the Blue Nile drought. However, when the correlation was calculated for the all years of ASO, it was −0.53. Previous studies also found high correlation during this season. When La Niña started in AMJ, JJA and JAS, in 67% of the cases there was a flood or extreme flood. There has to be an active event El Niño/La Niña during the season for development of the monsoon over Ethiopia (May–September), for this teleconnection to have an impact. We also found that in 67% of the cases in which El Niño was followed by La Niña there were extreme floods in the Blue Nile. The GPCP and UDEL rainfall anomalies showed the highest correlation with Niño 3.4 index during the AMJ through ASO season compared to the Blue Nile flow at Eldiem station, except for the CRU data set. So, there is potential to use the precipitation data set in the hydrological forecasting using information about ENSO.

An important conclusion is that JJAS rainfall in the upper catchment of the Blue Nile is highly sensitive to the Niño 3.4 SST anomaly during the early season of AMJ in Niño 3.4. This season is recommended by this study to be used in the seasonal forecasting of the Blue Nile. We also find that El Niño being immediately followed by La Niña conditions is predictive of extreme flood conditions in the upper Nile catchment, information that may also be useful in forecasting extreme floods over the region.

A version of this chapter first appeared as:

Zaroug, M., Eltahir, E. and Giorgi, F., 2014. Droughts and floods over the upper catchment of the Blue Nile and their connections to the timing of El Niño and La Niña events. Hydrology and Earth System Sciences, 18(3), pp.1239-1249.

Chapter 9—Simulating the connections of ENSO and the rainfall regime of East Africa and the upper Blue Nile region using a climate model of the Tropic

Modathir A. H. Zaroug
Elfatih A. B. Eltahir

The Blue Nile has a great impact on the life of millions of people in Ethiopia, Sudan and Egypt. It originates from Lake Tana and descends from the Ethiopian high plateau, with many tributaries in its upper course across the Ethiopian Highlands and two more tributaries (Dinder and the Rahad) in its lower course across Sudan. Although the Blue Nile is relatively short compared to the White Nile and it has a relatively small drainage area, it carries 60 to 70% of the total discharge and a great amount of sediment [Dumont, 2009].

In the study detailed in this chapter, we assessed whether these observed connections we have described through this book between precipitation in the East Africa/Blue Nile region and SSTs in the Pacific Ocean can be reproduced by using a physically based model of the climate system (the Regional Climate Model RegCM4 in

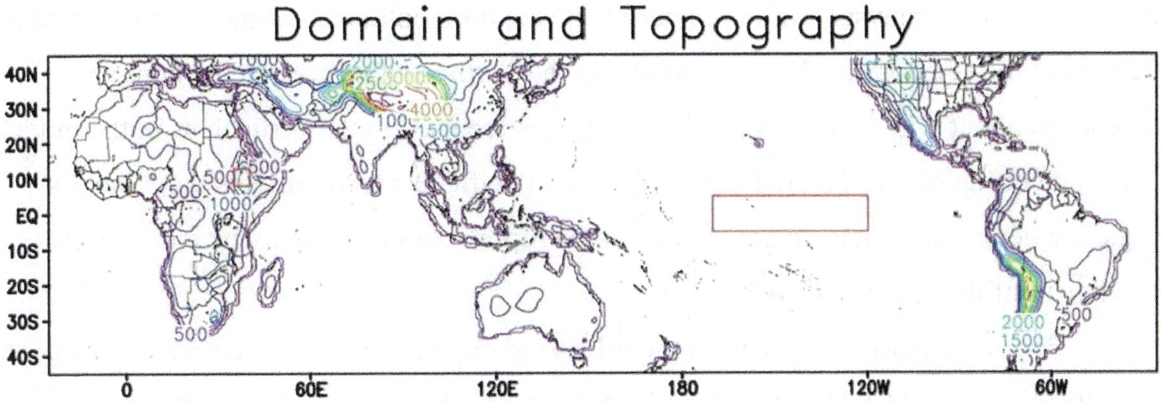

Figure 9.1: Fig. 1. The domain and the topography of the model. The red box in the Pacific Ocean shows the Niño 3.4 region, while the red box in the Ethiopian Highland shows the upper catchment of the Blue Nile (as in all following figures).

its tropical band configuration, Reg-TB, Giorgi et al., 2012; Coppola et al., 2012). Towards this purpose we completed and analyzed an ensemble of nine simulations of tropical climate driven by observed SSTs and north–south boundary conditions from the ERA-Interim reanalysis of observations for the 28-year period 1982–2009. In particular, the analysis addressed the impact of ENSO on the precipitation regime in Eastern Africa and the catchment of the Blue Nile, with a focus on the summer rainy season over the region.

RegCM4 description, experiments and data

One of the new developments of RegCM4 [Giorgi et al., 2012] is the tropical band configuration described by Coppola et al. [2012], or Reg-TB, by which the model domain covers the entire tropical belt with periodic boundary conditions in the x-direction and relaxation boundary conditions at the southern and northern boundaries. Figure 9.1 shows the model domain and topography along with some sub-regions selected for more detailed regional analysis. For our experiments the horizontal resolution is approximately 125 km at the equator, i.e. 360 grid points. In the north–south direction the domain extends to about ±45 degrees and 18 sigma layers are used in the vertical, as in the model standard configuration. The red rectangle in the Pacific Ocean represents the Niño 3.4 region, whereas the small square over Ethiopia represents our main focus area, i.e. the catchment of the upper Blue Nile.

For our experiments we used the following physics options described in Giorgi et al. [2012]: modified CCM3 radiative transfer scheme [Kiehl et al., 1996], modified planetary boundary layer scheme [Holtslag et al., 1990], SUBEX resolvable precipitation scheme [Pal et al., 2000], mixed cumulus convection configuration utilizing the scheme of Grell [1993] over land and that of Emanuel [1991] over oceans, and the biosphere–atmosphere transfer scheme [Dickinson et al., 1993] land surface package. As mentioned, the model used forcing lateral boundary conditions only in the northern and southern boundaries of the domain, with no external forcing in an east–west direction. The initial and lateral boundary conditions are provided by the ERA-Interim 1.5 ×1.5 gridded reanalysis [ERAI, Dee et al., 2011], which is the third-generation ECMWF reanalysis product. SSTs are from the weekly, 1 resolution Optimum Interpolation (OI) SST data set from the National Oceanic and Atmospheric Administration [Reynolds et al., 2007]. The simulation period is from 1 January 1982 to 31 March 2010 (28 years) and nine ensemble members are completed. The difference between the nine members is in the initial day of the simulation, which varies by 1 day. The first member starts on 1 January 1982, the second on 2 January 1982 and so on until the ninth member starts on 9 January 1982. The observed data mainly used for model validation are from the Global Precipitation Climatology Project (GPCP) version 2.2 [Huffman et al., 2011], available from January 1979 to December 2010 with a resolution of 2.5, although land data from the Climate Research Unit (CRU, land only, 0.5 ×0.5resolution [Mitchell et al., 2004]) is also used for more detailed regional analysis.

Finally, we point out that in our analysis we mostly focused on ensemble averaged results in order to filter out the internal variability of the model from the forced signal associated with the SST forcing [Shukla et al., 2000]. We however also show some examples of the spread across the ensemble members to provide a first-order assessment of this variability.

Validation of the model climatology over East Africa

In this section we first present a basic validation of the model climatology during the June–July–August–September (JJAS) rainy season over the broad East Africa region by comparing averages taken over the entire nine-member ensemble with available observations. Note that this model configuration is the same as that used by Coppola et al. [2012], and in that paper an assessment is provided of the model climatology over the entire tropical belt. In this regard, Coppola et al. [2012] show that, although regional biases in the model are present, the basic model climatology of equatorial and tropical regions is realistic. We also recall that, although we only analyze here data from the African region, the model was run for the full tropical band with forcing only at the northern and southern boundary, and therefore it is quite free to develop its own regional circulations and teleconnections within this large domain.

Rainfall

The spatial pattern of JJAS rainfall is compared to observations (both CRU and GPCP) in Figure 9.2. In JJAS the ITCZ is located in the Northern Hemisphere, therefore rainfall over the continent is mostly confined between 7°S and 18°N, while regions above and below these latitudes are relatively dry. This pattern of rainfall is mainly associated with the occurrence of mesoscale convective systems [d'Amato and Lebel, 1998; Jenkins et al., 2005]. Two rainfall maxima are located around the Cameroon Mountains and Ethiopian Highlands which are associated with local orographic features. The model captures the general patterns of the observed rainfall distribution, in particular the ITCZ position and intensity. However, rainfall over southern Sudan, central Africa and the Ethiopian Highlands is overestimated. Also, the monsoon rain belt appears narrower in the model than in the observational data sets, as indicated by the negative precipitation biases north and south of the main rain belt. In general, the performance of our Reg-TB configuration appears to be in line with previous studies performed using the RegCM system in various configurations and domains [Sun et al., 1999; Pal et al., 2007; Anyah and Semazzi, 2007; Sylla et al., 2010b; Steiner et al., 2009; Zaroug et al., 2013] or with other regional modelling systems [e.g., Vizy and Cook, 2002; Nikulin et al., 2012; Paeth et al., 2005; Gallée et al., 2004; Flaounas et al., 2010; Druyan et al., 2008].

Temperature

The seasonal average of JJAS 2m temperature for 1982 to 2009 is compared to CRU observations in Figure 9.3. The lowest temperatures are found mostly over the mountainous areas of Cameroon and Ethiopian Highlands, Tanzania and South Kenya,

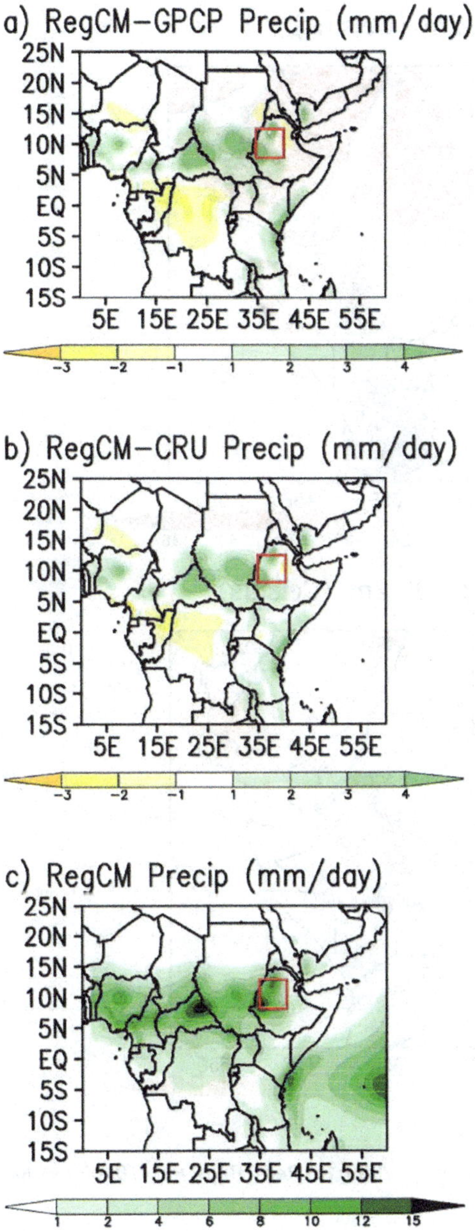

Figure 9.2: Averaged precipitation (in mm day−1) for JJAS 1982–2009: (a) Reg-TB minus CRU, (b) Reg-TB minus GPCP, and (c) RegCM. Model data are ensemble averages.

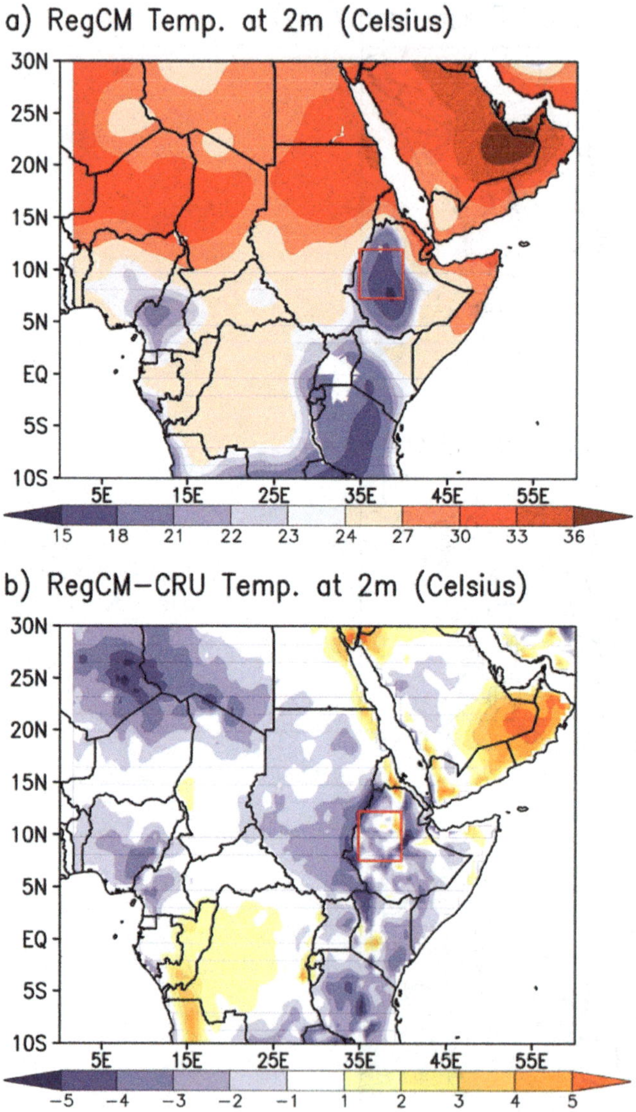

Figure 9.3: Averaged 2m air temperature (in °C) for JJAS 1982– 2009: (a) Reg-TB, (b) Reg-TB minus CRU. Model data are ensemble averages.

while the warmest areas are confined between 15°N and 27°N, with larger values over the Sahara. Reg-TB (Figure 9.3a) reproduces this spatial pattern but it shows a systematic cold bias of a few degrees in the convective regions of East Africa (Figure 9.3b), Nigeria, Algeria and Libya. This cold bias over tropical and equatorial Africa has been a persistent feature in RegCM, as also found for example in the experiments of Sylla et al. [2010b] and Coppola et al. [2012], although the magnitude of the bias is somewhat reduced in our simulation. It should be stressed that the CRU observations are possibly affected by large uncertainties in this region due to the lack of observing stations, particularly in remote and mountainous areas [Mitchell et al., 2004]. In addition, surface temperature depends on many parameters, including the presence of dust and aerosols. Given these uncertainties, we assessed that a model bias of a few degrees was acceptable for this study.

Outgoing Long-Wave Radiation (OLR)

Figure 4 compares simulated and observed (NOAA) averaged JJAS OLR, which is essentially a measure of cloudiness. Both the model result shown in Figure 9.4a and the observation result (not shown here) exhibit larger OLR values in North Africa because of small amounts of cloud cover, with correspondingly low OLR over the ITCZ. The model tends to overestimate the OLR over the Congo Basin (Figure 9.4b) as a result of an underestimation of cloudiness over this region, a result which is in line with previous applications of the model [Zaroug et al., 2013; Sylla et al., 2010a, b]. Slight overestimates are found over the Sahara and the Greater Horn of Africa (Fig Figure 9.4b), but in general

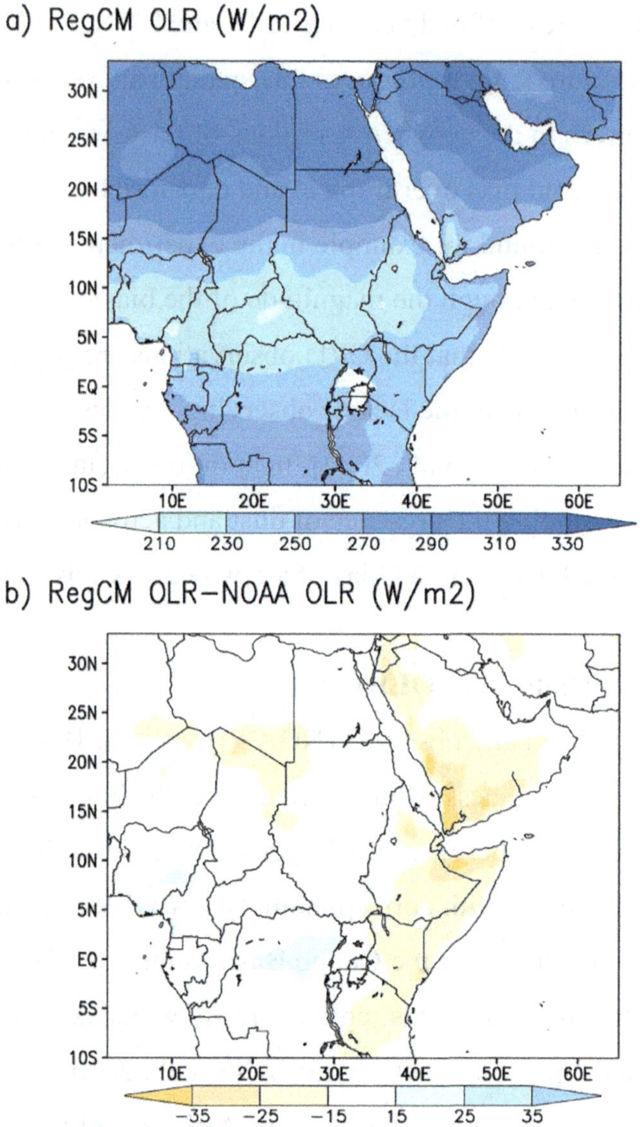

Figure 9.4: Averaged outgoing long-wave radiation (in Wm−2) for JJAS 1982–2009: (a) Reg-TB OLR, (b) Reg-TB OLR minus NOAA OLR. Model data are ensemble averages.

the model captures well the general observed features of the OLR pattern in both magnitude and spatial extent.

Low-level circulation, Tropical Easterly Jet and African Easterly Jet

The spatial patterns of average JJAS low level (925 mb) circulation are shown in Figure 9.5a for the ERA Interim reanalysis and Figure 9.5b for the Reg-TB simulation. Direct observations of the low-level wind over the region are not available, and thus we use here the reanalysis product for evaluating the model. Overall, the model reproduces well the main features of the mean low-level circulation, such as the low-level mon- soon flow over North Africa, the northerly Harmattan flow over East Africa, and the southeasterly flow over the Horn of Africa.

The Tropical Easterly Jet (TEJ), which affects the north African climate variability [Chen and Van Loon, 1987], develops between 200 and 150 mb in the upper

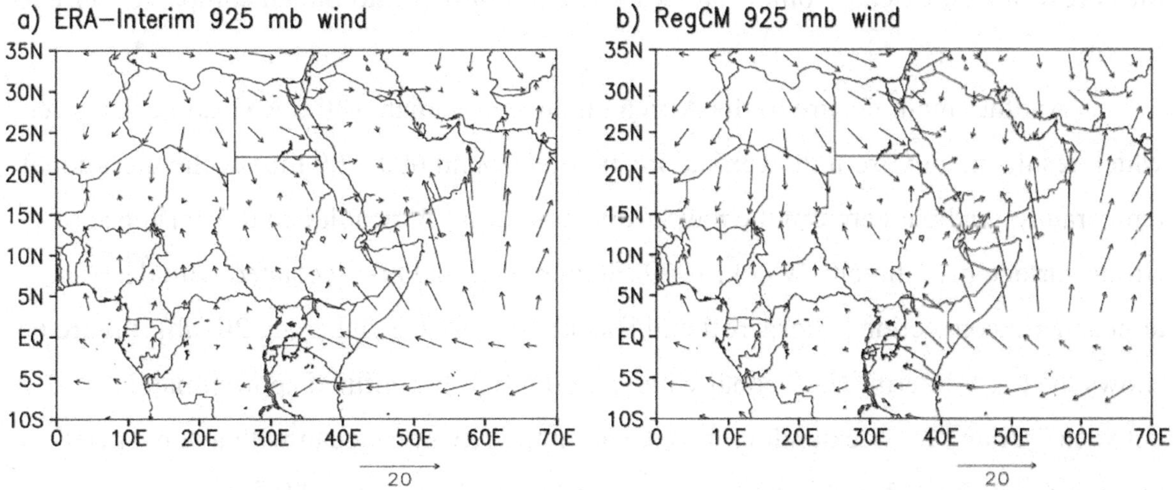

*Figure 9.5: Averaged 925 mb wind vector for JJAS 1982–2009: **(a)** ERA-Interim, **(b)** Reg-TB (ensemble average).*

troposphere over India in response to a large meridional thermal gradient and settles during the northern summer Asia monsoon season between the Tibetan Highlands and the Indian Ocean [Fontaine and Janicot, 1992; Koteswaram, 1958; Chen and van Loon, 1987]. It stretches from the Indochina peninsula, across the African continent, to the tropical Atlantic [Wu et al., 2009], and it is linked to anomalous SSTs on a planetary scale [Chen and van Loon, 1987].

Figure 9.6 shows that the TEJ in the ERA Interim reanalysis is confined between 3 and 17°N with a core speed exceeding $15ms^{-1}$. The band of the jet decreases gradually from East Africa to West Africa. In fact, the highest wind speed of about $18ms^{-1}$ occurs over the Horn of Africa and the western Indian Ocean, while the lowest speeds of about $6ms^{-1}$ are found over Niger. Reg-TB reproduces well the structure of the TEJ shown in the ERA Interim reanalysis, capturing both the location and intensity of the jet, although the core of the jet extends somewhat westward and further into Sudan compared to ERA-Interim.

Another main feature of the African monsoon climate is the African Easterly Jet, which results mainly from the vertical inversion (around 600–700 mb) of the meridional temperature gradient between the Sahara and equatorial Africa due to the existence of strong surface baroclinicity [Cook, 1999; Steiner et al., 2009] associated with deep atmospheric convection [Thorncroft and Blackburn, 1999; Sylla et al., 2010b]. Figure 6c shows the ERA-Interim JJAS zonal wind at 600 mb. It is confined approximately between 7°N and 20°N extending from Chad to the Atlantic Ocean with a core speed ranging from 11 to $13ms^{-1}$ located over West Africa. Again, Reg-TB simulates

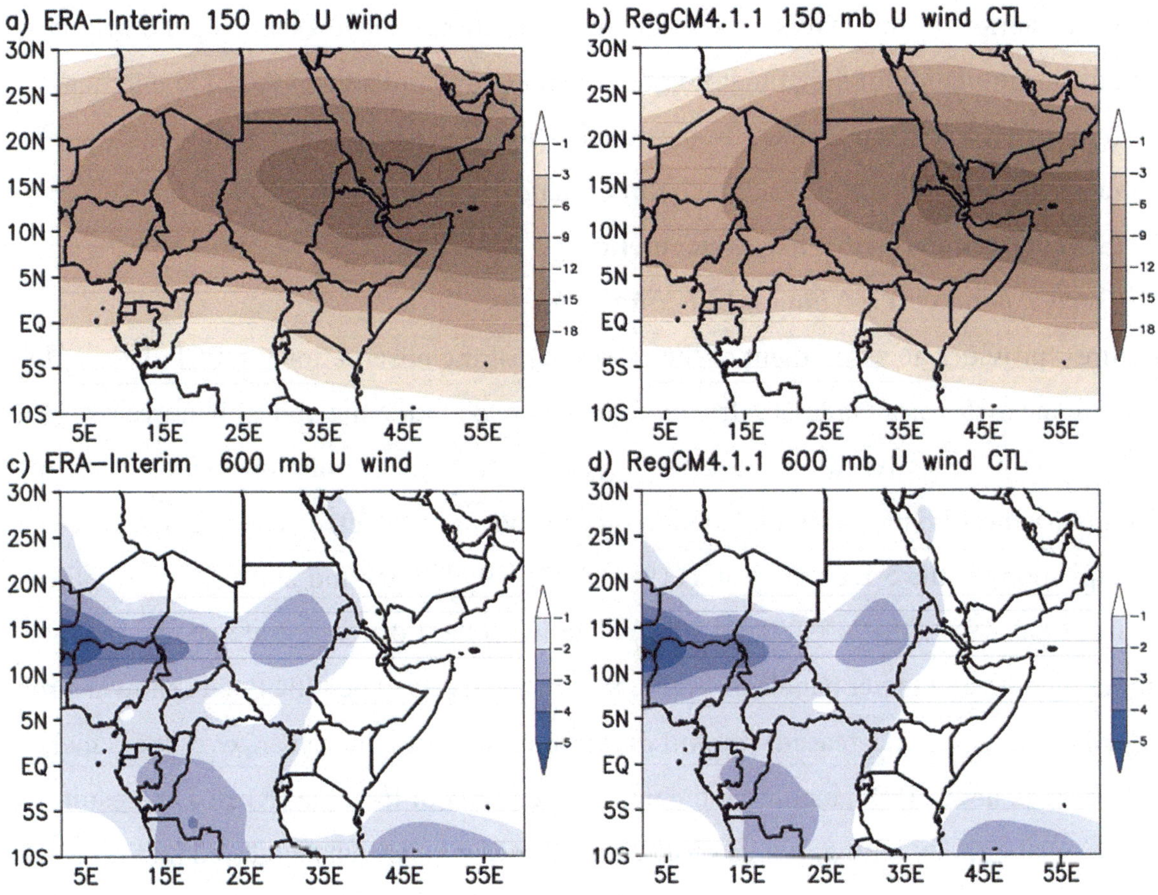

Figure 9.6: Averaged zonal wind (inms−1) for JJAS 1982–2009: (a) ERA Interim at 150 mb, (b) Reg-TB at 150 mb, (c) ERA Interim at 600 mb, and (d) Reg-TB at 600 mb. Model data are ensemble averages.

reasonably well both the strength and location of the African Easterly Jet (AEJ) (Figure 9.6d). In summary, the analysis presented in this section indicates that Reg-TB produces a realistic simulation of the climatology of the main dynamical and thermodynamical features of East African climate. In the next section we can thus turn our attention to the connections between ENSO and the hydroclimate of the upper Nile River Basin.

ENSO connections with East African climate

Difference between La Niña and El Niño years

Figure 7 provides an assessment of the capability of the observational GPCP data set to capture the differences in East African climate between La Niña and El Niño years. The GPCP data were analyzed for the same 28 years used in the simulations, 1982–2009, and 5 La Niña and El Niño years included in this period were selected. The average precipitation for the 5 La Niña years (1988, 1998, 1999, 2007 and 2008), the 5 El Niño years (1982, 1983, 1987, 1992 and 2002) along with their difference are shown in the upper, middle and lower panels of Figure 9.7, respectively. The GPCP data show a strong signal of increased (decreased) rainfall during La Niña (El Niño) years over the Sahel region up to about 18 N, including the upper catchment of the Nile River. This result agrees qualitatively with previous observational analyses, which suggested that La Niña years are associated with above normal rainfall and El Niño years with below normal rainfall in the upper catchment of the Blue Nile (Chapters 2 and 4; Amarasekera et al., 1997; De Putter et al., 1998; Camberlin et al., 2001; Abtew et al., 2009].

Figure 8 shows the corresponding fields averaged over the nine-member ensemble of Reg-TB simulations. The nine members were first averaged for each month of the 28

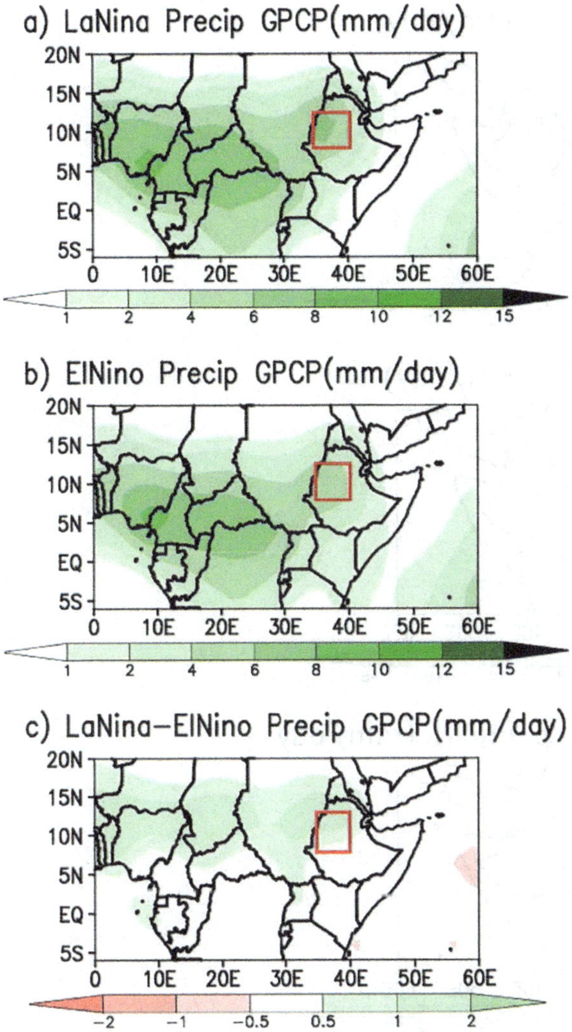

Figure 9.7: GPCP rainfall during JJAS for (a) 5 La Niña years (1988, 1998, 1999, 2007 and 2008), (b) 5 El Niño years (1982, 1983, 1987, 1992 and 2002), and (c) the difference between La Niña years and El Niño years.

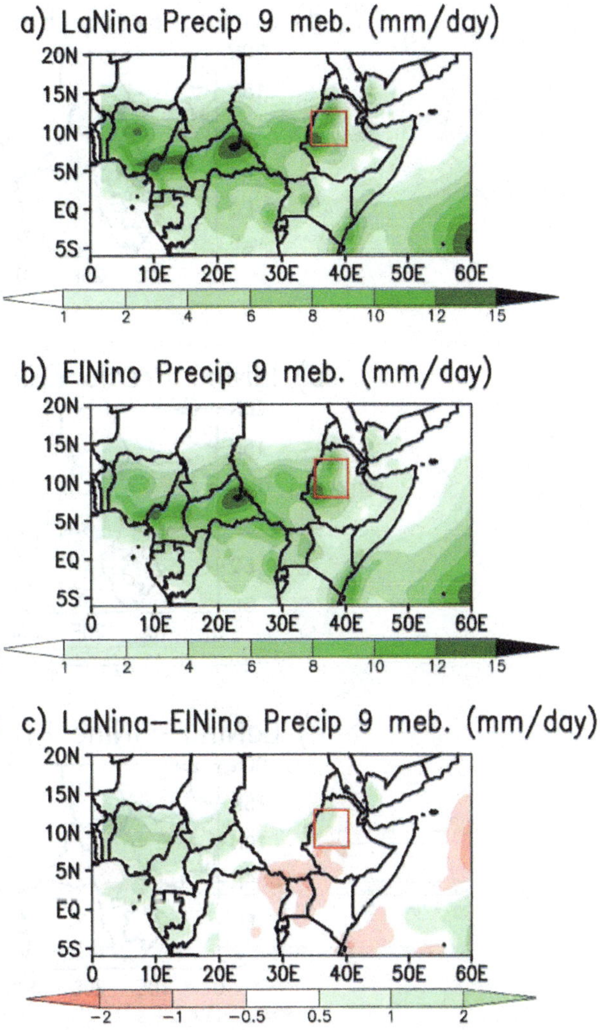

Figure 9.8: Ensemble average Reg-TB JJAS rainfall for (a) 5 La Niña years (1988, 1998, 1999, 2007 and 2008), (b) 5 El Niño years (1982, 1983, 1987, 1992 and 2002), and (c) the difference between La Niña years and El Niño years.

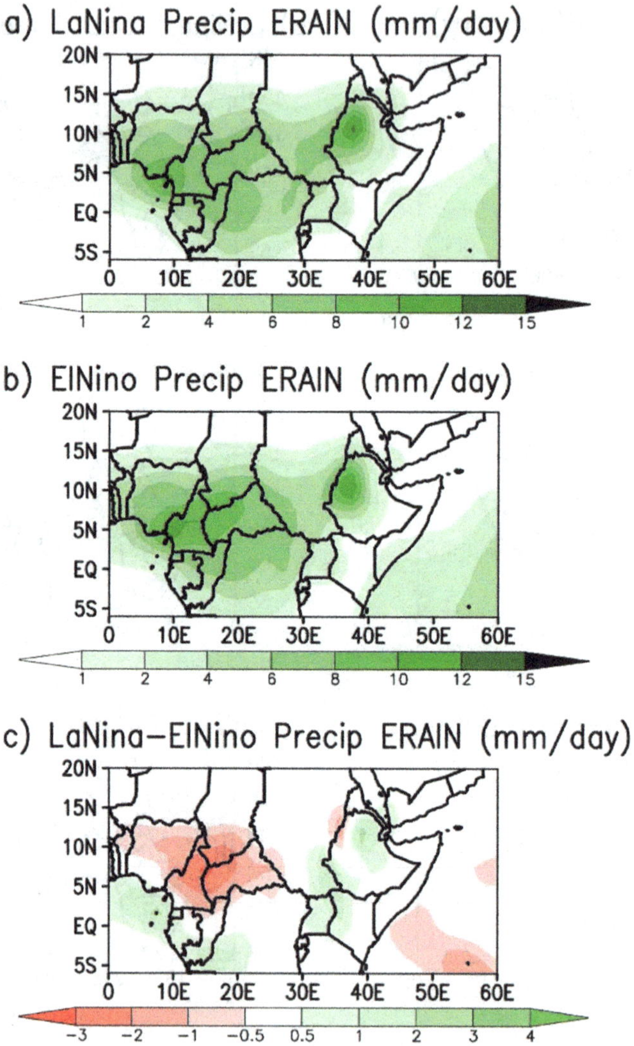

Figure 9.9: ERA Interim JJAS rainfall for (a) 5 La Niña years (1988, 1998, 1999, 2007 and 2008), (b) 5 El Niño years (1982, 1983, 1987, 1992 and 2002), and (c) the difference between La Niña years and El Niño years.

simulated years (1982–2009), and then the 5 La Niña and El Niño years were selected as in the observational analysis. The rainfall was then averaged for the rainy season (JJAS) and for the 5 La Niña and El Niño years to obtain ensemble-averaged values. It can be seen that the model captures the positive La Niña minus El Niño precipitation signal over the Sahel region, although this is less intense than in the observations and does not extend as far north. In addition, the model shows a negative signal in the region just North of Lake Victoria. The results of each ensemble member were also analyzed (not shown) and they showed a substantial variability across different members, which results in the average signal over the Sahel being weakened. Notwithstanding this problem, the model captured the ENSO-forced main positive signal over the Sahel band and the upper Nile catchment. We also calculated the same El Niño and La Niña precipitation signals (i.e. using the same 5 El Niño and La Niña years as in Figs. 7 and 8) from the ERA-Interim reanalysis (Figure 9.9). Surprisingly, we find a pronounced negative La Niña minus El Niño precipitation signal over the Sahel, in disagreement with the GPCP observations, along with shifted rainfall maxima in the upper Blue Nile area. We thus concluded that the internal physics of the Reg-TB corrected the representation of the teleconnection between ENSO and African precipitation found in ERA-Interim.

Correlation between rainfall anomalies over the Ethiopian highlands and SST anomalies in the Niño 3.4 region

Figure 9.10 shows the correlation between the ensemble average rainfall anomalies over the upper Nile river catchment during JJAS and the SST anomalies in the Niño 3.4 region

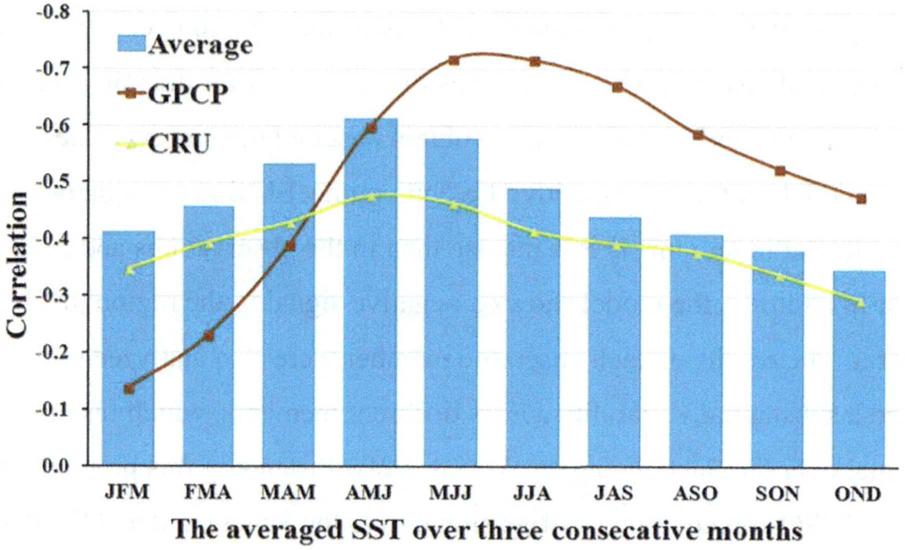

Figure 9.10: Correlation between rainfall anomalies over the Ethiopian Highlands (Reg-TB ensemble average, GPCP and CRU) and SST anomalies over the Pacific Ocean in the Niño 3.4 region.

during 3-month periods from January–February–March (JFM) to October–November–December (OND) calculated over a period of 28 years (1982 to 2009) using the equation

$$\mathrm{Correl}(X,Y) = \frac{\sum(x-\overline{x})(y-\overline{y})}{\sqrt{\sum(x-\overline{x})^2\sum(y-\overline{y})^2}} \qquad (9.1)$$

where $\overline{x}$ is the mean of the variable x (i.e. Niño 3.4 anomalies) and $\overline{y}$ is the mean of variable y (i.e. rainfall anomalies over the Ethiopia Highlands). The correlations are negative for all SST seasons, with the highest value of −0.62 during AMJ and the lowest of −0.35 in OND. Figure 9.10 also shows the same SST–precipitation correlations calculated for the GPCP and CRU observation data sets for the same 28 years. We first noted the substantial difference across the two data sets, with GPCP shifting the correlation maximum to the MJJ period, which confirms the uncertainties found in

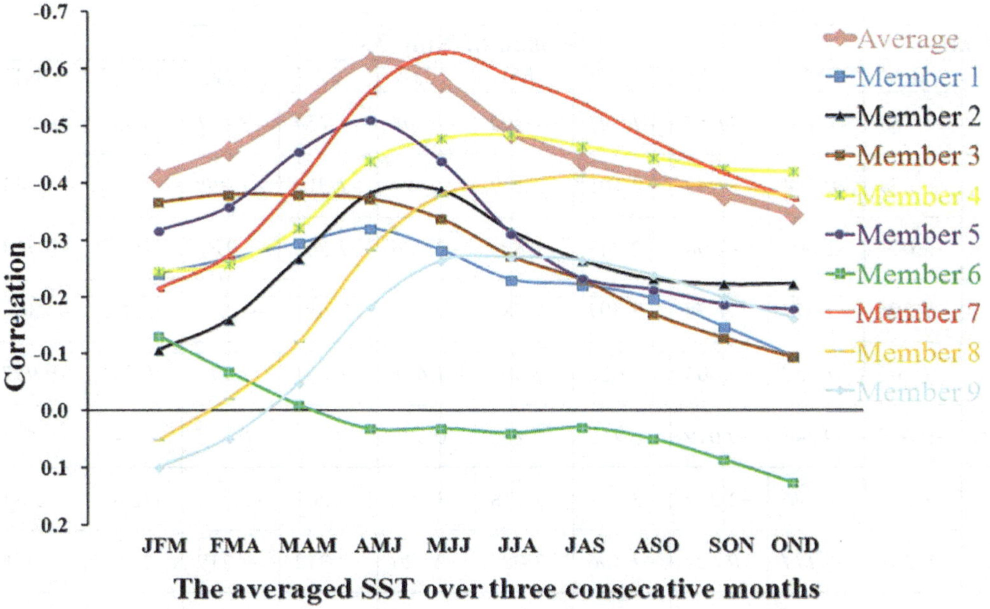

Figure 9.11: Correlation between rainfall anomalies over the Ethiopian Highlands (Reg-TB ensemble average and each individual ensemble member) and SST anomalies over the Pacific Ocean in the Niño 3.4 region.

precipitation observations over this region [e.g. Sylla et al., 2013]. The model correlations are in good agreement with the CRU ones, both in terms of magnitude and timing, thereby indicating a good performance by the model despite the observational uncertainties.

Figure 9.11 shows the corresponding SST–precipitation correlations (for the period 1982–2009) for each ensemble member. We found a considerable spread across members in both magnitude and timing of the correlation maximum. Most members, however, show values generally in line with the observational uncertainty with the prominent exception of member 6, which is very far from observations. An important conclusion from this analysis was that the internal variability of the tropical band model

Rainfall	**Season of Niño 3.4**									
	JFM	FMA	MAM	AMJ	MJJ	JJA	JAS	ASO	SON	OND
M1	*-1.268*	*-1.411*	*-1.57*	*-1.721*	*-1.502*	*-1.204*	*-1.154*	*-1.021*	*-0.752*	*-0.482*
M2	*-0.54*	*-0.816*	*-1.42*	-2.101	-2.139	*-1.694*	*-1.394*	*-1.208*	*-1.158*	*-1.165*
M3	*-2.004*	-2.092	-2.08	-2.043	*-1.825*	*-1.436*	*-1.201*	*-0.87*	*-0.65*	*-0.473*
M4	*-1.296*	*-1.374*	*-1.74*	-2.497	-2.786	-2.834	-2.681	-2.537	-2.399	-2.367
M5	*-1.703*	*-1.954*	-2.603	-3.022	-2.487	*-1.671*	*-1.221*	*-1.115*	*-0.976*	*-0.927*
M6	*-0.673*	*-0.348*	*-0.049*	*0.165*	*0.159*	*0.205*	*0.154*	*0.257*	*0.444*	*0.653*
M7	*-1.119*	*-1.446*	-2.229	-3.441	-4.118	-3.692	-3.262	-2.733	-2.34	-2.04
M8	*0.263*	*-0.114*	*-0.624*	*-1.506*	-2.088	-2.226	-2.312	-2.219	-2.201	-2.071
M9	*0.508*	*0.241*	*-0.246*	*-0.951*	*-1.399*	*-1.435*	*-1.406*	*-1.259*	*-1.042*	*-0.833*
Av.	-2.306	-2.627	-3.204	-3.962	-3.607	-2.858	-2.498	-2.292	-2.095	*-1.887*
Italic: below the critical value of –2.056 (not significant) Non italic: above the critical value of –2.056 (significant)										

Table 9.1: 95% significance test of the correlation between Niño 3.4 anomalies and rainfall at the upper catchment of the Blue Nile (see Figure 9.1) for the nine simulation members and for the ensemble average. The correlation is calculated over the period 1982–2009.

is high and that ensembles of simulations are needed to capture the atmospheric response to ENSO [Shukla et al., 2000]. Table 9.1 shows the significance of the correlations at the 95% confidence level, which was calculated for each ensemble member and for the average of the nine members. Members 1, 6 and 9 were below the 95% confidence level for all the seasons. The correlation is not significant for nearly all the members in JFM and FMA, a result in line with a recent study by Zaroug et al. [2014]. The average of the

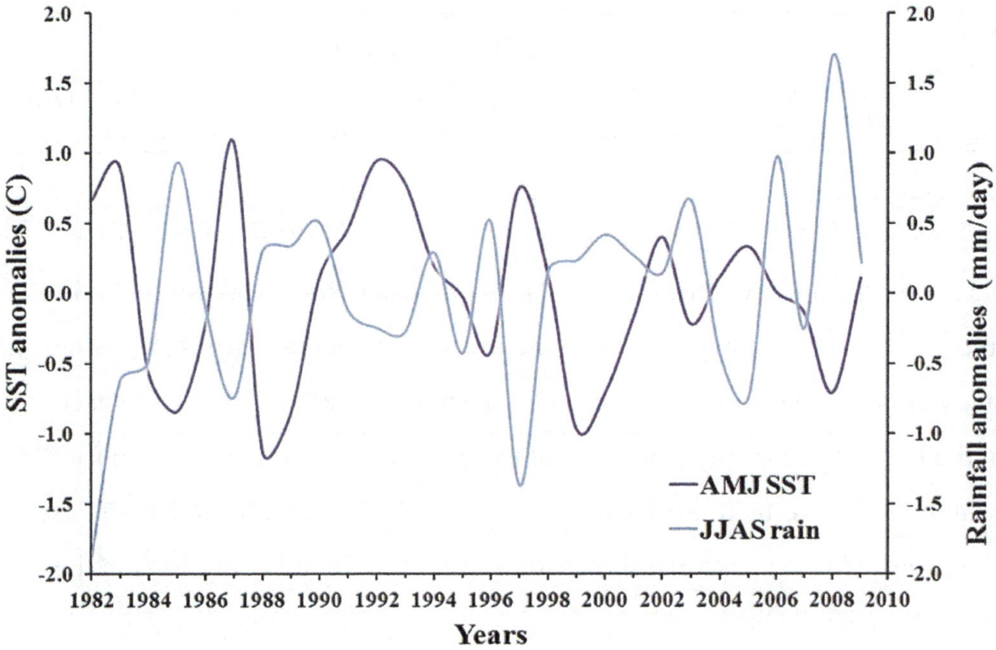

Figure 9.12: SST anomalies during AMJ in the Niño 3.4 region and ensemble average Reg-TB rainfall anomalies in the upper catchment of the Blue Nile.

nine members shows a highly significant correlation when compared with each ensemble member. The confidence test for the GPCP and CRU were calculated in Zaroug et al. [2014], who confirmed the strong impact of ENSO anomalies on the rainfall of the Ethiopia Highlands. Finally, Figure 9.12 shows the time evolution of AMJ SST anomalies and ensemble average JJAS precipitation over the upper Blue Nile river catchment, further evidencing the negative correlation between these two variables. The coefficient of determination (R2) for the linear fit between Niño 3.4 anomalies in different seasons and the JJAS averaged simulated rainfall anomalies at the upper catchment of the Blue Nile for the period 1982–2009 is shown in Table 9.2. AMJ shows

Rainfall anomalies	**Season of Niño 3.4 anomalies**									
	JFM	FMA	MAM	AMJ	MJJ	JJA	JAS	ASO	SON	OND
Average	0.17	0.21	0.28	0.38	0.33	0.24	0.19	0.17	0.14	0.12

Table 9.2: Coefficient of determination (R2) for the linear fit between Niño 3.4 in different seasons and the JJAS averaged simulated rainfall anomalies at the upper catchment of the Blue Nile for the period 1982–2009.

the highest coefficient of determination. The same calculation was completed for each ensemble member (not shown here) and member 7 showed the largest R^2 values. In the 2014 study detailed in Chapter 8, Zaroug et al. showed that the Blue Nile discharge showed the highest correlation for drought events with the Niño 3.4 SST during AMJ, a result which appears to be in line with Figure 9.10. These results support the use of Niño 3.4 SST during AMJ for seasonal forecasting of hydroclimate conditions in the upper catchment of the Blue Nile.

Regression analysis of Niño 3.4 index onto summer rainfall over the Sahel and East Africa

We investigated further the relationship between ENSO and summer rainfall over the Sahel through a regression analysis between SST in the Niño 3.4 region and rainfall in the summer for both the model and the GPCP observations. Figure 9.13 presents the regression of the JJA Niño 3.4 index onto the JJA GPCP rainfall from 1982 to 2009. The term **b** of the linear regression ($y = a + bx + e$) is calculated using the equation

$$b = \frac{\sum(x_i - \overline{x})(y_1 - \overline{y})}{\sqrt{\sum(x_i - \overline{x})^2}} \qquad (9.2)$$

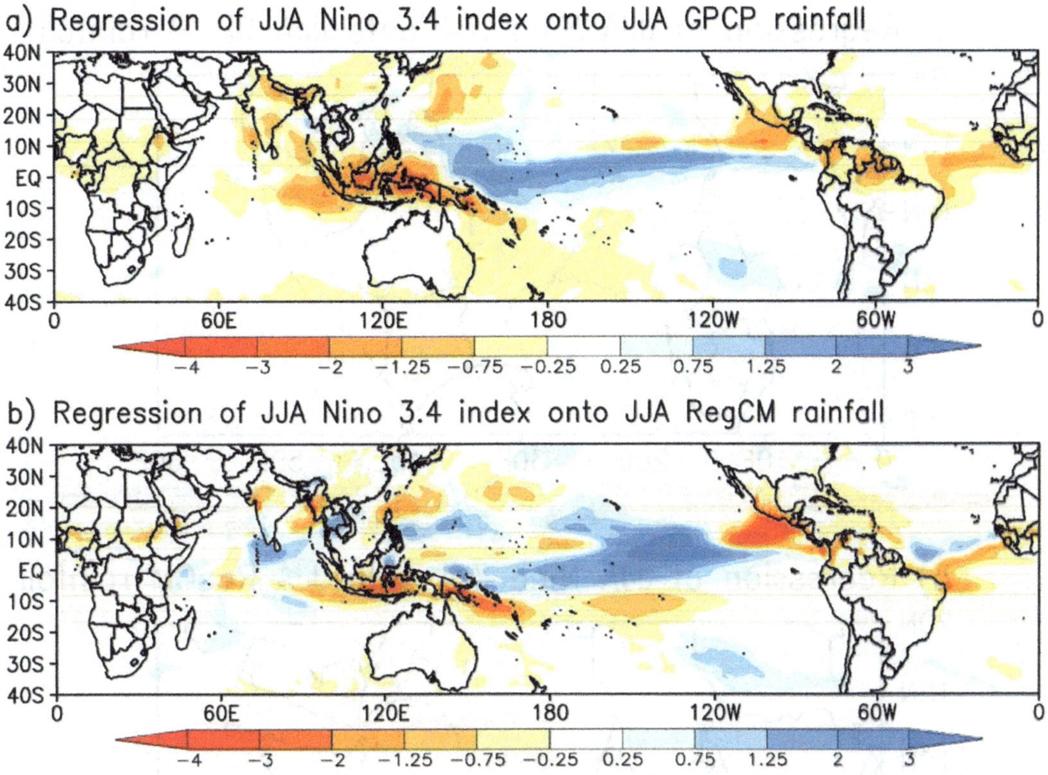

Figure 9.13: Regression of JJA Niño 3.4 index onto JJA rainfall for 1982–2009. (a) GPCP; (b) Reg-TB ensemble average.

where $\overline{x}$ is the mean of the variable x (i.e. Niño 3.4 anomalies) and $\overline{y}$ is the mean of variable y (i.e. rainfall anomalies). Niño 3.4 has an impact on many regions around the world, with a high positive signal in the equatorial Pacific and a negative signal over India, Southeast Asia and the northern part of South America. Figure 9.14 zooms over the Sahel and East Africa region and in the observations shows a negative signal over most of the Sahel and equatorial Africa regions, with a maximum over the Ethiopia Highlands.

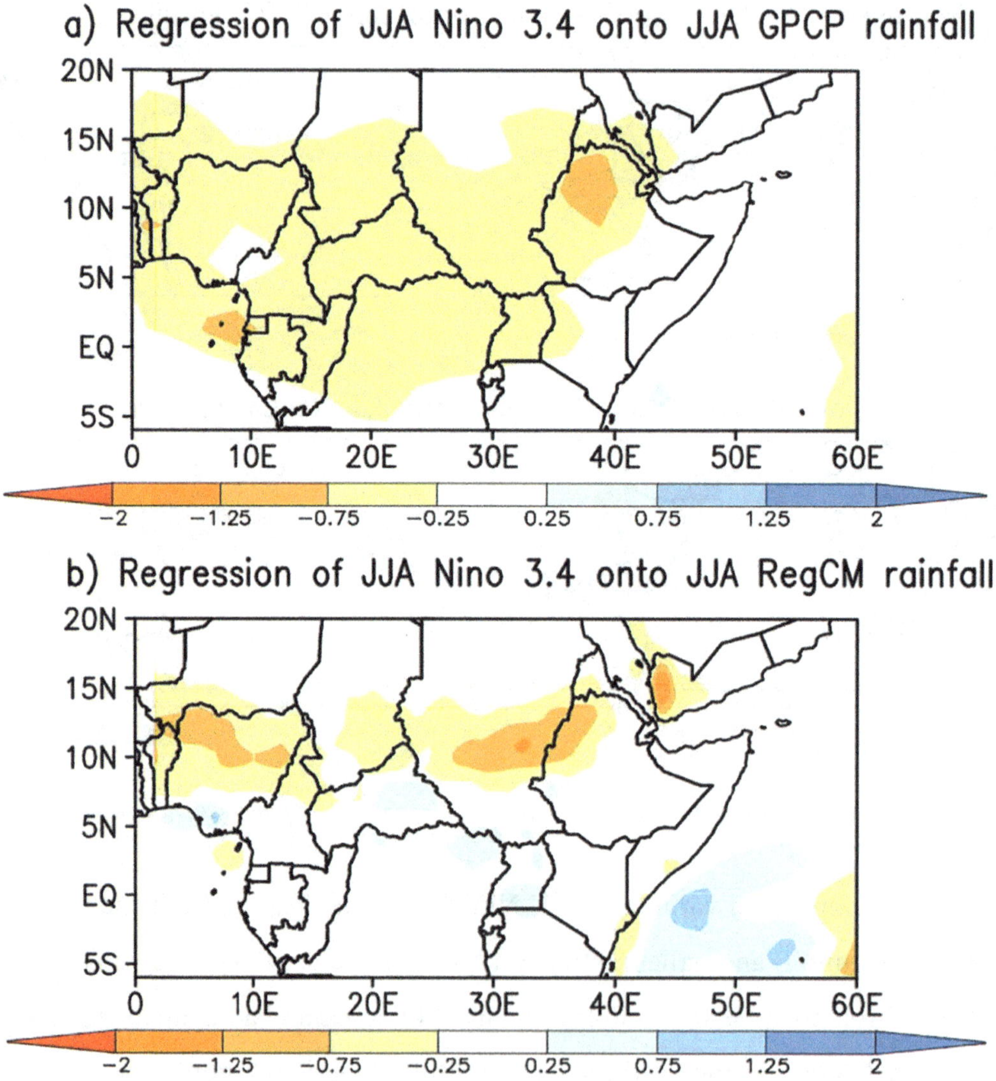

Figure 9.14: As Fig. 13 but for the Sahel and equatorial Africa region: (a) GPCP; (b) Reg-TB ensemble average.

The same regression analysis for the model ensemble average precipitation shows that, by and large, it reproduces the main features of the observed signal. Specifically, it reproduces the negative signal over the Sahel and East Africa region. Over equatorial Africa and the Congo Basin, the model shows areas of small positive signal not found in the observations. However, as already mentioned, observed precipitation over those areas may be characterized by large uncertainties due to lack of observing stations.

Conclusions

In this study we used the tropical band version of the RegCM4 system [Giorgi et al., 2012; Coppola et al., 2012] to simulate the observed statistical relationship between ENSO and the summer rainfall regime of the Sahel and East Africa, with a focus on the upper Blue Nile basin. Towards this objective an ensemble of nine 28-year long simulations was completed for a domain covering the entire tropical belt between 45 S and 45 N, driven by observed SST and north– south boundary conditions from the ERA-Interim reanalysis.

The climatology of Reg-TB over the Sahel and East Africa region was first evaluated against observations and the reanalysis product. It was shown that the model performs reasonably well in reproducing the observed climatology of temperature, rainfall, outgoing long-wave radiation and large-scale atmospheric circulation features. For example, the model captured well the rain belt, as well as the peaks over the Ethiopian, Guinea and Cameroon Highlands. In fact, this simulation even outperformed in some aspects previous applications of this model over the region [Sylla et al., 2010b,

Zaroug et al., 2013]. In addition, the lower-level and largescale circulation features affecting the monsoon (TEJ, AEJ) were realistically captured.

We then analyzed the ability of the model to reproduce the observed teleconnections between the Pacific SST and summer precipitation over the Sahel, East Africa and upper Nile River catchment. The model (average of nine members) was able to reproduce the observed negative correlation between the Niño 3.4 SST and precipitation over the focus regions, even outperforming the ERA-Interim reanalysis. It also showed the highest correlation of SST and summer precipitation over the Ethiopia Highlands during AMJ (−0.62), in line with previous observational studies. This result suggests that AMJ SSTs over the Niño 3.4 region can be a useful predictor in seasonal (JJAS) precipitation forecasting over East Africa and the Blue Nile region.

We finally investigated the SST–precipitation correlations in the individual ensemble members and found a wide spread across the ensemble members, which however reproduced the basic negative correlation pattern (except for one ensemble member). This confirms the need to perform ensembles of simulations to capture robust ENSO teleconnection signals (Shukla et al., 2000). Overall, our analysis provides encouraging indications towards the use of the Reg-TB configuration in seasonal forecasting, climate change and teleconnection studies over East Africa region.

A version of this chapter first appeared as:

Zaroug, M., Giorgi, F., Coppola, E., Abdo, G. and Eltahir, E., 2014. Simulating the connections of ENSO and the rainfall regime of East Africa and the upper Blue Nile region using a climate model of the Tropics. Hydrology and Earth System Sciences, 18(11), pp.4311-4323.

References

Chapter 2

1) Bhattacharya, R., Gupta, V. and Waymire, E. (1983). The Hurst effect under trends. *Journal of Applied Probability*, 20(03), pp.649-662.
2) Bliss, E. (2007). The Nile flood and world weather. *Quarterly Journal of the Royal Meteorological Society*, 53(221), pp.41-44.
3) Boes, D. and Salas, J. (1978). Nonstationarity of the mean and the hurst Phenomenon. *Water Resources Research*, 14(1), pp.135-143.
4) Bras, R. and Rodriguez-Iturbe, I. (1993). Random functions and hydrology. New York, NY: Dover Publ, p.559.
5) Cane, M., Zebiak, S. and Dolan, S. (1986). Experimental forecasts of El Niño. *Nature*, 321(6073), pp.827-832.
6) Hurst, H. (1951). Long-term storage capacity of reservoirs. *Trans. Am. Soc. Civ. Eng.,* 116, pp.770-808.
7) Janowiak, J. (1988). An Investigation of Interannual Rainfall Variability in Africa. *Journal of Climate*, 1(3), pp.240-255.
8) Klemeš, V. (1974). The Hurst Phenomenon: A puzzle?. *Water Resources Research*, 10(4), pp.675-688.
9) Moss, M., Pearson, C. and McKerchar, A. (1994). The Southern Oscillation index as a predictor of the probability of low streamflows in New Zealand. *Water Resources Research,* 30(10), pp.2717-2723.
10) Potter, K. (1976). Evidence for nonstationarity as a physical explanation of the Hurst Phenomenon. *Water Resources Research*, 12(5), pp.1047-1052.
11) Quinn, W. H., (1992) *El Niño: Historical and Paeoclimatic Aspects of the Southern Oscillation*, edited by H. F. Diaz and V. Markgraf, pp. 119-149, Cambridge Univ. Press, New York.
12) Rasmusson, E. and Carpenter, T. (1983). Variations in Tropical Sea Surface Temperature and Surface Wind Fields Associated with the Southern Oscillation/El Niño. *Monthly Weather Review*, 111, pp.517-528.
13) Ropelewski, C. and Halpert, M. (1987). Global and Regional Scale Precipitation Patterns Associated with the El Niño/Southern Oscillation. *Monthly Weather Review*, 115(8), pp.1606-1626.
14) Salas, J., Boes, D., Yevjevich, V. and Pegram, G. (1979). Hurst phenomenon as a pre-asymptotic behavior. *Journal of Hydrology*, 44(1-2), pp.1-15.
15) Simpson, H., Cane, M., Lin, S., Zebiak, S. and Herczeg, A. (1993). Forecasting Annual Discharge of River Murray, Australia, from a Geophysical Model of ENSO. *Journal of Climate*, 6(2), pp.386-390.
16) Trenberth, K. and Shea, D. (1987). On the evolution of the Southern Oscillation. *Monthly Weather Review*, 115, pp.3078-3096.
17) Wang, X. and Ropelewski, C. (1995). An Assessment of ENSO-Scale Secular Variability. *Journal of Climate*, 8(6), pp.1584-1599.

18) Wright, P. (1989). Homogenized long-period Southern Oscillation indices. *International Journal of Climatology*, 9(1), pp.33-54. (Received August 5, 1994; revised September 21, 1995; accepted September 22, 1995.)

Chapter 3

1) Allan R., J. Lindesay, and D. Parker, (1998) E1 Niño Southern Oscillation and Climatic Variability. 405pp., CSIRO Australia Publications.
2) Amarasekera, N. A., R. F. Lee, E. R. Williams, and E. A. B. Eltahir, (1997) ENSO and the natural variability in the flow of tropical rivers, *Journal of Hydrology*, 200, 24-39.
3) Anderson R. Y., (1992) Long-term changes in the frequency of occurrence of E1 Niño events. In *El Niño: Historical and Paleoclimatic Aspects of the Southern Oscillation*, edited by Diaz, H. F., and Markgraf, V., 193-200.
4) Bliss, E. W., (1925) The Nile flood and world weather, *Memoirs Roy. Meteorol. Soc.*, 4(23), 53-84.
5) Diaz, H. F., and R. S. Pulwarty, (1992) A comparison of Southern Oscillation and E1 Niño signals in the tropics. In *El Niño: Historical and Paleoclimatic Aspects of the Southern Oscillation*, edited by Diaz, H. F., and Markgraf, V., 175-192.
6) Eltahir, E. A. B., (1996) E1 Niño and the natural variability in the flow of the Nile river, *Water Resources Research*, 32 (1), 131-137.
7) Harrison, D. E., and N. K. Larkin, (1997) Darwin sea level pressure, 1876-1996: evidence for climate change?, *Geophysical Research Letters*, 24(14), 1779-1782.
8) Meehl, G. A., and W. M. Washington, (1996) E1 Niño-like climate change in a model with increased atmospheric C02 concentrations, *Nature*, 382, 56-60.
9) Popper, W., (1951) The Cairo Nilometer. 269pp., Berkeley and Los Angeles, University of California Press.
10) Quinn, W. H., (1992) a study of Southern Oscillation-related climatic activity for A.D. 622-1990 incorporating Nile River flood data, In *El Niño: Historical and Paleoclimatic Aspects of the Southern Oscillation*, edited by Diaz, H. F., and Markgraf, V., 119-149.
11) Rajagopalan, B., U. Lall, and M. Cane, (1997) Anomalous ENSO occurrences: an alternate view, *Journal of Climate*, 10(9), 2351-2357.
12) Tousson, O., (1925) Memoire sur l'Histoire du Nil, in Memoirres a L'Institut D'Egypte. Cairo. 544pp., Imprimerie de l'Institut Francaise d'Archeologie Orienta.
13) Trenberth, K. E., and T. J. Hoar, (1996) The 1990-1995 E1 Niño- Southern Oscillation event: longest on record Trenberth, Kevin E.- Hoar, Timothy J., *Geophysical Research Letters*, 23, 57-60.
14) Trenberth, K. E., T. J. Hoar, (1997) E1 Niño and climate change, *Geophysical Research Letters*, 24(23), 3057-3060,

15) Wang, G., and E. A. B. Eltahir, (1999) Use of ENSO information in medium- and long-range forecasting of the Nile floods, *Journal of Climate.*
16) Wright P. B., (1989) Homogenized long-period Southern Oscillation indices, *International Journal of Climatology*, 9, 33-54.

Chapter 4

1) Afifi, A. A., and S. P. Azen, (1979) *Statistical Analysis—A Computer Oriented Approach.* Academic Press, 442 pp.
2) Amarasekera, N. A., R. F. Lee, E. R. Williams, and E. A. B. Eltahir, (1997) ENSO and the natural variability in the flow of tropical rivers. *Journal of Hydrology.,* 200, 24–39.
3) Bliss, E. W., (1925) Nile flood and world weather. *Memoirs Roy. Meteorol. Soc.,* 4, 53–84.
4) Camberlin, P., (1995) June–September rainfall in north-eastern Africa and atmospheric signals over the tropics: A zonal perspective. *International Journal of Climatology.,* 15, 773–783.
5) Cane, M. A., S. E. Zebiak, and S. C. Dolan, (1986) Experimental forecasts of El Niño. *Nature,* 321, 827–832.
6) Dracup, J. A., and E. Kahya, (1994) The relationship between U.S. streamflow and La Niña events. *Water Resource Research,* 30, 2133– 2141.
7) Eltahir, E. A. B., (1996) El Niño and the natural variability in the flow of the Nile river. *Water Resources Research,* 32, 131–137.
8) Folland, C., J. Owen, M. N. Ward, and A. Colman, (1991) Prediction of seasonal rainfall in the Sahel region using empirical and dynamical methods. *Journal of Forecasting,* 10, 21–56.
9) Hulme, M., (1995) Estimating global changes in precipitation. *Weather,* 50, 34–42.
10) Hurst, H. E., (1957) *The Nile, A General Account of the River and the Utilization of Its Water.* Constable, 326 pp.
11) Kahya, E., and J. A. Dracup, (1993) U.S. streamflow patterns in relation to the El Niño/Southern Oscillation. *Water Resources Research,* 29, 2491–2503.
12) Kahya, E., and J. A. Dracup, (1994) The influence type 1 El Niño and La Niña events on streamflows in the Pacific southwest of the United States. *Journal of Climate,* 7, 965–976.
13) McBride, J. L., and N. Nicholls, (1983) Seasonal relationship between Australian rainfall and the Southern Oscillation. *Monthly Weather Review,* 111, 1998–2004.
14) Miller, R. G., (1962) *Statistical Prediction by Discriminant Analysis. Meteor. Monogr.,* No. 25, Amer. Meteor. Soc., 54 pp.
15) Parsons, R., (1978) *Statistical Analysis: A Decision-Making Approach.* Harper and Row, 791 pp.
16) Piechota, T. C., and J. A. Dracup, (1996) Drought and regional hydrologic variation in the United States: Association with the El Niño–Southern Oscillation. *Water Resources Research,* 32, 1359–1373.
17) Richey, J., C. Nobre, and C. Deser, (1989) Amazon River discharge and climate variability: 1903 to 1985. *Science,* 246, 101–103.
18) Ropelewski, C. F., and M. S. Halpert, (1987) Global and regional scale precipitation patterns associated with the El Niño/Southern Oscillation. *Monthly Weather Review,* 115, 1606–1626.

19) Ropelewski, C. F., and M. S. Halpert, (1989) Precipitation patterns associated with the highindex phase of the Southern Oscillation. *Journal of Climate,* 2, 268–284.
20) Simpson, H. J., M. A. Cane, A. L. Herczeg, S. E. Zebiak, and J. H. Simpson, (1993a) Annual river discharge in Southern Australia related to El Niño–Southern Oscillation forecasts of sea surface temperature. *Water Resources Research,* 29, 3671–3680.
21) Simpson, H. J., M. A. Cane, S. K. Lin, A. L. Herczeg, and S. E. Zebiak, (1993b) Forecasting annual discharge of River Murray, Australia, from a geophysical model of ENSO. *Journal of Climate,* 6, 386–390.
22) Singh, V. P., (1995) *Computer Models of Watershed Hydrology.* Water Resources Publications, 1130 pp.
23) Trenberth, K. E., and D. J. Shea, (1987) On the evolution of the Southern Oscillation. *Monthly Weather Review,* 115, 3078–3096.
24) West, M., (1989) *Bayesian Forecasting and Dynamic Models.* Springer, 704 pp.
25) Winkler, R., 1972: *An Introduction to Bayesian Inference and Decision.* Holt, Rinehart and Winston, 563 pp.
26) Wright, P. B., (1989) Homogenized long-period Southern Oscillation indices. *International Journal of Climatology.,* 9, 33–54.
27) Zebiak, S. E., and M. A. Cane, (1987) A model El Niño–Southern Oscillation. *Monthly Weather Review,* 115, 2262–2278.

Chapter 5

1) Baecher, G., Anderson, R., Britton, B., Brooks, K., Gaudet, J. (2015) The Nile Basin: Environmental Transboundary Opportunities and Constraints Analysis. Report for United States Agency for International Development.
2) Beyene, T., Lettenmaier, DP., Kabat, P. (2009) Hydrologic impacts of climate change on the Nile River Basin: implications of the 2007 IPC scenarios. *Climatic Change*, 100(3-4), 433-461.
3) Cai, W. et al. (2014) Increasing frequency of extreme El Niño events due to greenhouse warming. *Nature Climate Change* 4, 111–116.
4) Cai, W. et al., (2015) Increased frequency of extreme La Niña events under greenhouse warming. *Nature Climate Change* 5, 132-137.
5) Cervigni, R., Liden, R., Neumann, J.E. and Strzepek, K.M. eds. (2015) Enhancing the Climate Resilience of Africa's Infrastructure: The Power and Water Sectors. *World Bank Publications.*
6) Christensen, J.H. et al. (2007) Regional Climate Projections. In: Climate Change 2007: *The Physical Science Basis. Contribution of Working Group I to the Fourth Assessment Report of the Intergovernmental Panel on Climate Change* [Solomon, S., D. Qin, M. Manning, Z. Chen, M. Marquis, K.B. Averyt, M. Tignor and H.L. Miller (eds.)]. Cambridge University Press, Cambridge, United Kingdom and New York, NY, USA.
7) Christensen, J.H. et al., (2013) Climate Phenomena and their Relevance for Future Regional Climate Change. In: *Climate Change 2013: The Physical Science Basis. Contribution of Working Group I to the Fifth Assessment Report of the Intergovernmental Panel on Climate Change* [Stocker, T.F., D. Qin, G.-K. Plattner, M. Tignor, S.K. Allen, J. Boschung, A. Nauels, Y. Xia, V.

Bex and P.M. Midgley (eds.)]. Cambridge University Press, Cambridge, United Kingdom and New York, NY, USA, pp. 1217–1308, doi: 10.1017/CBO9781107415324.028.

8) Conway, D. (2005) From headwater tributaries to international river: observing and adapting to climate variability and change in the Nile basin. *Global Environmental Change*, 15(2), pp.99-114.
9) Conway, D. and Schipper, E.L.F. (2011) Adaptation to climate change in Africa: Challenges and opportunities identified from Ethiopia. *Global Environmental Change*, 21(1), pp.227-237.
10) Conway, D., Hulme M.. (1993) Recent Fluctuations in precipitation and runoff over the Nile subbasins and their impact on Main Nile discharge. *Climatic Change*, 25, 127 -151.
11) El Daw, A., Salas, J. D., Garcia, L. A. (2003) Long Range Forecasting of the Nile River Flows Using Climate Forcing. *Journal of Applied Meteorology*, 42:890-904.
12) El.Shamy, M. E. (2009) Impacts of climate change on Blue Nile flows using bias corrected GCM scenarios. *Hydrology and Earth System Sciences*, 13, 551–565.
13) Eltahir, E. A. B. (1996) El Niño and the natural variability in the flow of the Nile River. Water Resources Research 32(1):131–137.
14) FAO (Food and Agriculture Organization of the United Nations), (2000) New Dimensions in Water Security: Water, Society and Ecosystem Services in the 21st century. Rome; Food and Agriculture Organization of the United Nations.
15) Gebre, SL., Ludwig, F. (2015) Hydrological Response to Climate Change of the Upper Blue Nile River Basin: Based on IPCC Fifth Assessment Report (AR5). *Journal of Climatolgy and Weather Forecasting* 3: 121. doi:10.4172/2332-2594.1000121
16) Hurst, H. E., (1951) Long term storage capacity of reservoirs, *Trans. Am. Soc. Civ. Eng.*, 116, 770–779.
17) Hurst, H. E., Black, R. P., Simaika Y. M,. (1965) Long-Term Storage: An Experimental Study, Constable, London.
18) Kim, S. T. et al. (2014) Response of El Niño sea surface temperature variability to greenhouse warming, *Nature Climate Change*, 4, 786–790.
19) Kim, U., Kaluarachchi, J. J. (2000) Climate change impacts on water resources in the Upper Blue Nile River Basin, Ethiopia. *Journal of the American Water Resources Association*, 45, 6.
20) King, A. and Block, P. (2014) An assessment of reservoir filling policies for the Grand Ethiopian Renaissance Dam. *Journal of Water and Climate Change*, 5(2), pp.233-243.
21) Rayner NA, et al, (2003) Global analyses of sea surface temperature, sea ice, and night marine air temperature since the late nineteenth century. *Journal of Geophysical Research* 108(D14):4407
22) Siam, M. S., Eltahir, E. A. B. (2015) Explaining and forecasting interannual variability in the flow of the Nile River, *Hydrology of Earth Systems Science*, 19, 1181-1192, doi:10.5194/hess-19-1181-2015.
23) Siam, M. S., Wang, G., Demory, M. E., Eltahir, E. A. B. (2014) Role of the Indian Ocean sea surface temperature in shaping the natural variability in the flow of Nile River. *Climate Dynamics* 43 (3-4): 1011-1023.
24) Taye, M. T., Ntegeka, V., Ogiramoi, N. P., Willems, P. (2011) Assessment of climate change impact on hydrological extremes in two source regions of the Nile River Basin, *Hydrology of*

Earth Systems Science, 15, 209-222, doi:10.5194/hess-15-209-2011.

25) Taylor, KE., Stouffer, RJ., Meehl, GA, (2015). An overview of CMIP5 and the experiment design. *Bulletin of the American Meteorological Society* 93:485–498.
26) UNEP. (2013) Adaptation to Climate-change Induced Water Stress in the Nile Basin: A Vulnerability Assessment Report. Division of Early Warning and Assessment (DEWA). United Nations Environment Programme (UNEP). Nairobi, Kenya.
27) United Nations, Department of Economic and Social Affairs, Population Division. (2013) World Population Prospects: The 2012 Revision, Highlights and Advance Tables. Working Paper No. ESA/P/WP.228.
28) Vörösmarty, C.J., Fekete, B.M. and Tucker, B.A. (1998). Global River Discharge, 1807–1991, Version. 1.1 (RivDIS). Data set. Available on-line [http://www. daac. ornl. gov] from Oak Ridge National Laboratory Distributed Active Archive Center, Oak Ridge, Tennessee, USA.
29) Vose, R.S., R.L. Schmoyer, P.M. Steurer, T.C. Peterson, R. Heim, T.R. Karl, and J.K. Eischeid. (2016). Global Historical Climatology Network, 1753-1990. ORNL DAAC, Oak Ridge, Tennessee, USA. http://dx.doi.org/10.3334/ORNLDAAC/220
30) Wang, G., Eltahir, E. A. B. (1999) Use of ENSO information in Medium and Long Range Forecasting of the Nile Floods, *Journal of Climate*, 12, 1726-1737.
31) Yamana, T.K., 2015. Mechanistic modelling of the links between environment, mosquitoes and malaria transmission in the current and future climates of West Africa (Doctoral dissertation, Massachusetts Institute of Technology).
32) Yates, D. N., Strzepek, K. M. (1998) An assessment of integrated climate change impacts on the agricultural economy of Egypt. *Climatic Change*, 38(3): 261- 287.

Chapter 6

1) Abtew, W., Melesse, A. M., and Dessalegne, T. (2009) El Niño Southern Oscillation link to the Blue Nile River Basin hydrology, *Hydrological Processes*, 23, 3653–3660, doi:10.1002/hyp.7367.
2) Amarasekera, K. N., Lee, R. F., Williams, E. R., and Eltahir, E. A. B. (1996) ENSO and the natural variability in the flow of tropical rivers, *Journal of Hydrology*, 200, 24–39.
3) Bacmeister, J. T., Pegion, P. J., Schubert, S. D., and Suarez, M. J. (2000) An atlas of seasonal means simulated by the NSIPP 1 atmospheric GCM, NASA Tech. Memo. 104606, Vol. 17, Goddard Space Flight Center, Greenbelt, 194 pp.
4) Beltrando, G. and Camberlin, P. (1993) Interannual variability of rainfall in Eastern Horn of Africa and indicators of atmospheric circulation, *International Journal of Climatology*, 13, 533–546.
5) Black, E., Slingo, J., and Sperber, K. R. (2003) An observational study of the relationship between excessively strong short rains in coastal East Africa and Indian Ocean SST, *Monthly Weather Review*, 31, 74– 94.
6) Camberlin, P. (1995) June–September rainfall in North-Eastern Africa and atmospheric signals over the tropics: a zonal perspective, *International Journal of Climatology*, 15, 773–783.
7) Conway, D. and Hulme, M. (1993) Recent Fluctuations in precipitation and runoff over the Nile subbasins and their impact on Main Nile discharge, *Climatic Change*, 25, 127–151.

8) ElDaw, A., Salas, J. D., and Garcia, L. A. (2003) Long Range Forecasting of the Nile River Flows Using Climate Forcing, *Journal of Applied Meteorology*, 42, 890–904.
9) ElSanabary, M. H., Gan, T. Y., and Mwale, D. (2014) Application of wavelet empirical orthogonal function analysis to investigate the nonstationary character of Ethiopian rainfall and its teleconnection to nonstationary global Sea Surface Temperature variations for 1900–1998, *International Journal of Climatology,* 34, 1798–1813, doi:10.1002/joc.3802.
10) Eltahir, E. A. B. (1996) El Niño and the natural variability in the flow of the Nile river, *Water Resources Research,* 32, 131–137.
11) Giro, D., Grimes, D., and Black, E. (2011) Teleconnections between Ethiopian summer rainfall and sea surface temperature: part I Observation and Modeling, *Climate Dynamics*, 37, 103–199.
12) Gissila, T., Black, E., Grimes, D. I. F., and Slingo, J. M. (2004) Seasonal forecasting of the Ethiopian Summer rains, *International Journal of Climatology*, 24, 1345–1358.
13) Graham, R., Gordon, M., McLean, P. J., Ineson, S., Huddleston, M. R., Davey, M. K., Brookshaw, A., and Barnes, R. T. H. (2005) A performance comparison of coupled and uncoupled versions of the Met Office seasonal prediction general circulation model, *Tellus A*, 57, 320–339.
14) Hastenrath, S., Polzin, D., and Mutai, C. (2011) Circulation Mechanisms of Kenya Rainfall Anomalies, *Journal of Climate*, 24, 404–412.
15) Ju, J. and Slingo, J. M. (1995) The Asian summer monsoon and ENSO, *Quarterly Journal of the Royal Meteorological Society*, 121, 1133–1168.
16) Kawamura, R. (1998) A possible mechanism of the Asian summer monsoon–ENSO coupling, *Journal of the Meteorological Society of Japan.,* 76, 1009– 1027.
17) Melesse, A., Abtew, W., Setegn, S. G., and Desalegn, T. (2011) Hydrological Variability and Climate of the Upper Blue Nile River Basin, in: *Nile River Basin: Hydrology, Climate and Water Use*, 1st Edn., edited by: Melesse, A. M., Springer, New York, USA, 3–37, doi:10.1007/978-94-007-0689-7_1.
18) Molteni, F., Stockdale, T., Balmaseda, M., Balsamo, G., Buizza, R., Ferranti, L., Magnusson, L., Mogensen, K., Palmer, T., and Vitart, F.(2011) The new ECMWF seasonal forecast system (System 4), ECMWF Research Department Technical Memorandum no. 656, ECMWF, Reading, UK, 51 pp.
19) Penland, C. and Matrosova, L. (1998) Prediction of Tropical Atlantic Sea Surface Temperatures Using Linear Inverse Modeling, *Journal of Climate*, 11, 483–496.
20) Rayner, N. A., Parker, D. E., Horton, E. B., Folland, C. K., Alexander, L. V., Rowell, D. P., Kent, E. C., and Kaplan, A. (2003) Global analyses of sea surface temperature, sea ice, and night marine air temperature since the late nineteenth century, *Journal of Geophysical Research*, 108, 4407, doi:10.1029/2002JD002670.
21) Saha, S., Moorthi, S., Pan, H.-L., Wu, X., Wang, J., Nadiga, S., Tripp, P., Kistler, R., Woollen, J., Behringer, D., Liu, H., Stokes, D., Grumbine, R., Gayno, G., Wang, J., Hou, Y.-T., Chuang, H.-Y., Juang, H.-M. H., Sela, J., Iredell, M., Treadon, R., Kleist, D., Van Delst, P., Keyser, D., Derber, J., Ek, M., Meng, J., Wei, H., Yang, R., Lord, S., van den Dool, H., Kumar, A., Wang, W., Long, C., Chelliah, M., Xue, Y., Huang, B., Schemm, J.-K., Ebisuzaki, W., Lin, R., Xie, P., Chen,

M., Zhou, S., Higgins, W., Zou, C.-Z., Liu, Q., Chen, Y., Han, Y., Cucurull, L., Reynolds, R. W., Rutledge, G., and Goldberg, M. (2010) The NCEP Climate Forecast System Reanalysis, *Bulletin of the American Meteorological Society*, 91, 1015–1057, doi:10.1175/2010BAMS3001.1.

22) Saha, S., Moorthi, S., Wu, X., Wang, J., Nadiga, S., Tripp, P., Pan, H.-L., Behringer, D., Hou, Y.-T., Chuang, H.-Y., Iredell, M., Ek, M., Meng, J., Yang, R., van den Dool, H., Zhang, Q., Wang, W., and Chen, M. (2013) The NCEP Climate Forecast System Version 2, *Journal of Climate*, 27, 2185–2208.
23) Shukla, J. and Wallace, J. M. (1983) Numerical simulation of the atmospheric response to equatorial Pacific sea surface temperature anomalies, *Journal of the Atmospheric Sciences*, 40, 1613–1630.
24) Siam, M. S., Wang, G., Estelle, M. E., and Eltahir, E. A. B. (2014) Role of the Indian Ocean Sea Surface Temperature in shaping natural variability in the flow of the Nile River, *Climate Dynamics*, 43, 1011– 1023.
25) Soman, M. K. and Slingo, J. (1997) Sensitivity of the Asian summer monsoon to aspects of sea-surface-temperature anomalies in the tropical pacific ocean, *Quarterly Journal of the Royal Meteorological Society,* 123, 309-336,.
26) Tomczak, M. and Godfrey, J. S. (1995) *Regional oceanography: an introduction*, Pregamon, London.
27) Trenberth, K. E. (1997) The Definition of El Niño, *Bulletin of the American Meteorological Society*, 78, 2771–2777.
28) Vörösmarty, C. J., Fekete, B., and Tucker, B. A. (1998) River discharge database, Version 1.1 (RivDIS v 1.0 supplement), available through the institute for the study of earth, oceans and space University of New Hampshire, Durham, NH, USA.
29) Wang, G. and Eltahir, E. A. B. (1999) Use of ENSO information in Medium and Long Range Forecasting of the Nile Floods, *Journal of Climate*, 12, 1726–1737.
30) West, M. (1989) Bayesian Forecasting and Dynamic Models, *Springer*, New York, 704 pp.
31) Winkler, R. (1972) *An introduction to Bayesian inference and Decision*, Holt, Rinchart and Winstoon, New York, 563 pp.
32) Xue, Y., Leetmaa, A., and Ji, M. (2000) ENSO prediction with Markov model: The impact of sea level, *Journal of Climate*, 13, 849–871.
33) Yang, J., Liu, Q., Xie, S.-P., Liu, Z., and Wu, L. (2007) Impact of the Indian Ocean SST basin mode on the Asian summer monsoon, *Geophysical Research Letters*, 34, L02708, doi:10.1029/2006GL028571.

Chapter 7

1) Amarasekera K.N., Lee R.F., Williams E.R., Eltahir E.A.B. (1996) ENSO and the natural variability in the flow of tropical rivers. *Journal of Hydrology* 200:24–39
2) Beltrando G., Camberlin P. (1993) Interannual variability of rainfall in Eastern Horn of Africa and indicators of atmospheric circulation. *International Journal of Climatology* 13:533–546
3) Black E., Slingo J., Sperber K.R. (2003) An observational study of the relationship between excessively strong short rains in coastal East Africa and Indian Ocean SST. *Monthly Weather*

Review 31:74–94
4) Camberlin P. (1995) June–September rainfall in North-Eastern Africa and atmospheric signals over the tropics: a zonal perspective. *International Journal of Climatology* 15:773–783
5) Camberlin P. (1997) Rainfall anomalies in the source region of the Nile and their connection with the Indian summer monsoon. *Journal of Climate* 10:1380–1392
6) Cole J.E. et al (2000) Tropical pacific forcing of decadal SST variability in the western Indian Ocean over the past two centuries. *Science* 287(5453):617–619
7) Conway D., Hulme M. (1993) Recent fluctuations in precipitation and runoff over the Nile sub-basins and their impact on main Nile discharge. *Climate Change* 25:127–151
8) Dee D.P. et al (2011) The ERA-Interim reanalysis: configuration and performance of the data assimilation system. *Quarterly Journal of the Royal Meteorological Society* 137:553–597
9) ElDaw A., Salas J.D., Garcia L.A. (2003) Long-range forecasting of the Nile River flows using climate forcing. *Journal of Applied Meteorology* 42: 890–904
10) ElShamy M.E. (2009) Impacts of climate change on Blue Nile flows using bias corrected GCM scenarios. *Hydrology and Earth Systems Science* 13:551–565
11) Eltahir E.A.B. (1996) El Nino and the natural variability in the flow of the Nile River. *Water Resour Research* 32(1):131–137
12) Giannini A., Saravanan R., Chang P. (2003) Oceanic forcing of Sahel rainfall on interannual to interdecadal time scales. *Science* 302:1027–1030
13) Hastenrath S., Polzin D., Mutai C. (2011) Circulation mechanisms of Kenya rainfall anomalies. *Journal of Climate* 24:404–412
14) IPCC (2007) The physical science basis. Contribution of working group I to the fourth assessment report of the intergovernmental panel on climate change. In: Solomon S., Qin D., Manning M., Chen Z., Marquis M., Averyt K.B., Tignor M., Miller H.L. (eds) *Climate change* 2007. Cambridge University Press, Cambridge, United Kingdom and New York, NY, USA
15) Jobson J.D. (1991) Applied multivariate data analysis. *Volume I: regression and experimental design*. Springer, New York, pp 182–185
16) Johns T.C. et al (2006) The new hadley centre climate model (HadGEM1): evaluation of coupled simulations. *Journal of Climate* 19: 1327–1353
17) Ju J., Slingo J.M. (1995) The Asian summer monsoon and ENSO. *Quarterly Journal of the Royal Meteorological Society* 121:1133–1168
18) Kalnay E. et al (1996) The NCEP/NCAR 40-year reanalysis project. *Bulletin of the American Meteorological Society* 77:437–471
19) Kawamura R. (1998) A possible mechanism of the Asian summer monsoon-ENSO coupling. *Journal of the Meteorological Society of Japan* 76:1009–1027
20) Martin G.M., Ringer M.A., Pope V.D., Jones A., Dearden C., Hinton T.J. (2006) The physical properties of the atmosphere in the New Hadley Centre Global Environmental Model (HadGEM1). Part I: model description and global climatology. *Journal of Climate* 19(7):1274–1301. doi:10.1175/JCLI3636.1
21) Mitchell T.D., Jones P.D. (2005) An improved method of constructing a database of monthly

climate observations and associated highresolution grids. *International Journal of Climatology* 25:693–712

22) Mutai C., Neil Ward M. (2000) East African rainfall and the tropical circulation/convection on intraseasonal to interannual timescales. *Journal of Climate* 13:3915–3939
23) Nicholson Sharon E. (1997) An analysis of the ENSO signal in the tropical Atlantic and western Indian Oceans. *International Journal of Climatology* 17(4):345–375
24) Pohl B., Camberlin P. (2006) Influence of the Madden–Julian oscillation on East African rainfall. Part I: intraseasonal variability and regional dependency. *Quarterly Journal of the Royal Meteorological Society* 132:2521–2539
25) Rayner N.A., Parker D.E., Horton E.B., Folland C.K., Alexander L.V., Rowell D.P., Kent E.C., Kaplan A. (2003) Global analyses of sea surface temperature, sea ice, and night marine air temperature since the late nineteenth century. *Journal of Geophysical Research* 108(D14):4407
26) Ringer M.A., Martin G.M., Greeves C.Z., Hinton T.J., James P.M., Pope V.D., Scaife A.A., Stratton R.A., Inness P.M., Slingo J.M., Yang G.Y. (2006) The physical properties of the atmosphere in the New Hadley Centre Global Environmental Model (HadGEM1). Part II: aspects of variability and regional. *Journal of Climate* 19:1302–1326
27) Shukla J., Wallace J.M. (1983) Numerical simulation of the atmospheric response to equatorial Pacific sea surface temperature anomalies. *Journal of the Atmospheric Sciences* 40:1613–1630
28) Siam S.M., Demory M.E., Eltahir E.A.B. Hydrological cycles over the Congo and upper Blue Nile basins: evaluation of general circulation models simulations and reanalysis data products. *Journal of Climate*. doi:10.1175/JCLI-D-12-00404.1
29) Soman M.K., Slingo J. (1997) Sensitivity of the Asian summer monsoon to aspects of sea-surface-temperature anomalies in the tropical Pacific Ocean. *Quarterly Journal of the Royal Meteorological Society* 123:309–336
30) Taylor K.E., Williamson D., Zwiers F. (2000) The sea surface temperature and sea-ice concentration boundary conditions for AMIP II simulations. Technical Report 60, PCMDI
31) Tomczak M, Godfrey JS (1995) *Regional oceanography: an introduction*. Pregamon, London
32) Trenberth K.E. (1997) The definition of El Niño. *Bulletin of the American Meteorological Society* 78:2771–2777
33) Uppala SM et al (2005) The ERA-40 re-analysis. *Quarterly Journal of the Royal Meteorological Society* 131:2961–3012
34) Vörösmarty C.J., Fekete B., Tucker B.A. (1998) River discharge database, Version 1.1 (RivDIS v 1.0 supplement). Available through the institute for the study of earth, oceans and space University of New Hampshire, Durham NH (USA)
35) Wang G., Eltahir E.A.B. (1999) Use of ENSO information in medium and long range forecasting of the Nile floods. *Journal of Climate* 12: 1726–1737
36) Yang J.L., Liu Q.Y., Xie S.P. et al (2007) Impact of the Indian Ocean SST basin mode on the Asian summer monsoon. *Geophysical Research Letters* 34:L02708. doi:10.1029/2006GL028571

Chapter 8

1) Abtew, W., Melesse, A. M., and Dessalegne, T. (2009) El Niño Southern Oscillation link to the

Blue Nile River Basin hydrology, *Hydrological Processes*, 23, 3653–3660, doi:10.1002/hyp.7367.
2) Amarasekera, K. N., Lee, R. F., Williams, E. R., and Eltahir, E. A. B. (1997) ENSO and the natural variability in the flow of tropical rivers, *Journal of Hydrology*, 200, 24–39.
3) Awadalla, A. G. and Rousselle, J. (1999) Forecasting the Nile flood using sea surface temperatures as inputs: A comparison between transfer function with noise and neural networks, in: *Proc. 19th AGU Hydrology Days*, Fort Collins, CO, American Geophysical Union, 23–36.
4) Barnett, T. P., Graham, N. E., Cane, M. A., Zebiak, S. E., Dolan, S. C., O'Brien, J., and Legler, D. M. (1988) On the prediction of the El Niño of 1986–1987, *Science*, 241, 192–196.
5) Barnston, A. G., van den Dool, H. M., Rodenhuis, D. R., Ropelewski, C. R., Kousky, V. E., O'Lenic, E. A., and Leetmaa, A. (1994) Long-lead seasonal forecasts-Where do we stand?, *Bulletin of the American Meteorological Society*, 75, 2097–2114.
6) Bhatt, U. S. (1989) Circulation regimes of rainfall anomalies in the African–South Asian monsoon belt, *Journal of Climate*, 2, 1133–1144.
7) Camberlin, P. (1997) Rainfall anomalies in the source region of the Nile and their connection with the Indian summer monsoon, *Journal of Climate*, 10, 1380–1392.
8) Camberlin, P., Janicot, S., and Poccard, I. (2001) Seasonality and atmospheric dynamics of the teleconnection between African rainfall and tropical sea surface temperature: Atlantic vs. ENSO, *International Journal of Climatology*, 21, 973–1005.
9) Chen, D. and Cane, M. A. (2008) El Niño prediction and predictability, *Journal of Computational Physics*, 227, 3625–3640.
10) Chen, D., Cane, M. A., Kaplan, A., Zebiak, S. E., and Huang, D. (2004) Predictability of El Niño over the past 148 years, *Nature*, 428, 733–736.
11) Conway, D. (1997) A water balance model of the Upper Blue Nile in Ethiopia, *Hydrological Sciences Journal*, 42, 265–286.
12) Conway, D. (2000) The climate and hydrology of the Upper Blue Nile River, *Geographical Journal*, 166, 49–62.
13) De Putter, T., Loutre, M., and Wansard, G. (1998) Decadal periodicities of Nile River historical discharge (AD 622–1470) and climatic implications, *Geophysical Research Letters*, 25, 3193–3196.
14) Duethmann, D., Zimmer, J., Gafurov, A., Güntner, A., Kriegel, D., Merz, B., and Vorogushyn, S. (2013) Evaluation of areal precipitation estimates based on downscaled reanalysis and station data by hydrological modelling, *Hydrology and Earth Systems Sciences*, 17, 2415–2434, doi:10.5194/hess-17-2415-2013.
15) Dumont, H. J. (1986) The Nile River system, in: *The Ecology of River Systems*, edited by: Davies, B. R. and Walker, K. F., Kluwer, Dordrecht, 61–74.
16) Eldaw, A. K., Salas, J. D., and Garcia, L. A. (2003) Long-range forecasting of the Nile River flows using climatic forcing, *Journal of Applied Meteorology*, 42, 890–904.
17) Eltahir, E. A. B. (1996) El Niño and the natural variability in the flow of the Nile River, *Water Resources Research*, 32, 131–137.
18) Eltayeb, G. E. (2003) Khartoum, Sudan, UN-HABITAT Case Studies, London, 1–20. ESRL

(Earth System Research Laboratory), USDC (U.S. Department of Commerce), Physical Sciences Division: Data available at: http://www.esrl.noaa.gov/psd/, last access: August 2013, 1900–2010.
19) Gissila, T., Black, E., Grimes, D. I. F., and Slingo, J. M. (2004) Seasonal forecasting of the Ethiopian summer rains, *International Journal of Climatology*, 24, 1345–1358.
20) Gray, W. M., Christopher, W. L., Mielke Jr., P. W., and Berry, K. J. (1992) Predicting Atlantic seasonal hurricane activity 6–11 months in advance, *Weather and Forecasting*, 7, 440–455,.
21) Hastenrath, S., Polzin, D., and Camberlin, P. (2004) Exploring the predictability of the "short rains" at the coast of East Africa, *International Journal of Climatology*, 24, 1333–1343.
22) Huffman, G. J., Bolvin, D. T., and Adler, R. F. (2011) GPCP Version 2.2 Combined Precipitation Data set, WDC-A, NCDC, Asheville, NC, Data set, available at: http://www.ncdc.noaa.gov/oa/wmo/ wdcamet-ncdc.html (last access: October 2011).
23) Hulme, M. and Trilsbach, A. (1989) The August 1988 storm over Khartoum: its climatology and impact, *Weather*, 44, 82–90.
24) Jury, M. R. (2004) The coherent variability of African river flows: composite climate structure and the Atlantic circulation, *Water SA*, 29, 1–10.
25) Latif, M., Anderson, D., Barnett, T., Cane, M., Kleeman, R., Leetmaa, A., and Schneider, E. (1998) A review of the predictability and prediction of ENSO, *Journal of Geophysical Research*, 103, 14375–14393.
26) Mitchell, T. D., Carter, T. R., Jones, P. D., Hulme, M., and New, M. (2004) A comprehensive set of high-resolution grids of monthly climate for Europe and the globe: the observed record (1901–2000) and 16 scenarios (2001–2100), Tyndall Centre for Climate Change Research, Working Paper No. 55, 25 pp.
27) MOWRE (2011) Flood Report 2011, Ministry of Water Resources and Electricity, Khartoum, Sudan.
28) Mutai, C.,Ward, M., and Colman, A. (1998) Towards the prediction of the East Africa short rains based on sea surface temperature atmosphere coupling, *International Journal of Climatology*, 18, 975–997.
29) Mwale, D. and Gan, T. (2005) Wavelet analysis of variability, teleconnectivity, and predictability of September–November East Africa rainfall, *Journal of Applied Meteorology*, 44, 256–269.
30) NOAA (National Weather Service) (2014) Monitoring and Data, Center for Weather and Climate Prediction, Maryland, USA, available at: http://www.cpc.ncep.noaa.gov/products/analysis_ monitoring/ensostuff/ensoyears.shtml, last access: August 2013.
31) Philippon, N., Camberlin, P., and Fauchereau, N. (1991) Empirical predictability study of October–December east African rainfall, *Quarterly Journal of the Royal Meteorological Society*, 128, 2239–2256, 2002. Seleshi, Y.: Statistical analysis of the Ethiopian droughts in the XXth century based on monthly and yearly precipitation totals, M.S. thesis, Vrije Universiteit, Brussels, Belgium.
32) Seleshi, Y. and Zanke, U. (2004) Recent changes in rainfall and rainy days in Ethiopia, *International Journal of Climatology*, 24, 973–983.
33) Sutcliffe, J. V., Dugdale, G., and Milford, J. R. (1989) The Sudan floods of 1988, *Hydrological*

Sciences Journal, 34, 355–364.
34) Teklu, T., von Braun, J., and Zaki, E. (1991) Drought and Famine Relationships in Sudan: Policy and Implications, International Food Policy Research Institute (IFPRI), Washington, DC, USA, Research report No. 88, 140 pp.
35) Trenberth, K. E. (1997) The definition of El Niño, *Bulletin of the American Meteorological Society*, 78, 2771–2777.

Chapter 9

1) Abtew, W., Melesse, A. M., and Dessalegne, T. (2009) El Niño Southern Oscillation link to the Blue Nile River Basin hydrology, *Hydrological Processes*, 23, 3653–3660.
2) Amarasekera, K. N., Lee, R. F., Williams, E. R., and Eltahir, E. A. B. (1997) ENSO and the natural variability in the flow of tropical rivers, *Journal of Hydrology*, 200, 24–39.
3) Anyah, R. O. and Semazzi, F. H. M. (2007) Variability of East African rainfall based on multiyear RegCM3 simulations, *Interntaional Journal of Climatology*, 27, 357–371.
4) Camberlin, P., Janicot, S., and Poccard, I. (2001) Seasonality and atmospheric dynamics of the teleconnection between African rainfall and tropical sea surface temperature: Atlantic vs. ENSO, *Interntaional Journal of Climatology*, 21, 973–1005.
5) Chen, T. C. and van Loon, H. (1987) Interannual variation of the tropical easterly jet, *Monthly Weather Review*, 115, 1739–1759.
6) Cook, K. H. (1999) Generation of the African easterly jet and its role in determining West African precipitation, *Journal of Climate*, 12, 1165– 1184.
7) Coppola, E., Giorgi, F., Mariotti, L., and Bi, X. (2012) RegT-Band: A tropical band version of RegCM4, *Climate Research*, 52, 115–133.
8) D'amato, N. and Lebel, T. (1998) On the characteristics of the rainfall events in the Sahel with a view to the analysis of climatic variability, *Interntaional Journal of Climatology*, 18, 955–974.
9) Dee, D. P., Uppala, S. M., Simmons, A. J., Berrisford, P., Poli, P., Kobayashi, S., Andrae, U., Balmaseda, M. A., Balsamo, G., Bauer, P., Bechtold, P., Beljaars, A. C. M., van de Berg, L., Bidlot, J., Bormann, N., Delsol, C., Dragani, R., Fuentes, M., Geer, A. J., Haimberger, L., Healy, S. B., Hersbach, H., Hólm, E. V., Isaksen, L., Kållberg, P., Köhler, M., Matricardi, M., McNally, A. P., Monge-Sanz, B. M., Morcrette, J. J., Park, B. K., Peubey, C., de Rosnay, P., Tavolato, C., Thépaut, J. N., and Vitart, F. (2011) The ERA-Interim reanalysis: configuration and performance of the data assimilation system, *Quarterly Journal of the Royal Meteorological Society*, 137, 553–597, doi:10.1002/qj.828.
10) De Putter, T., Loutre, M., and Wansard, G. (1998) Decadal periodicities of Nile River historical discharge (AD 622–1470) and climatic implications, *Geophysical Research Letters*, 25, 3193–3196.
11) Diaz, H. F., Hoerling, M. P., and Eischeid, J. K. (2001) ENSO variability, teleconnections and climate change, *Interntaional Journal of Climatology*, 21, 1845– 1862.
12) Dickinson, R. E., Henderson-Sellers, A., Kennedy, P. J., and Wilson, M. F. (1993) Biosphere–atmosphere transfer scheme (BATS) version 1e as coupled to the NCAR Community Climate Model, NCAR Tech. Note, NCAR/TN387+ STR.

13) Druyan, L. M., Fulakeza, M., and Lonergan, P. (2008) The impact of vertical resolution on regional model simulation of the west African summer monsoon, *Interntaional Journal of Climatology*, 28, 1293–1314.
14) Dumont, H. J. (2009) *The Nile: origin, environments, limnology and human use*, Monog. Biol., Vol. 89, Springer Netherlands.
15) Elagib, N. A. and Elhag, M. M. (2011) Major climate indicators of ongoing drought in Sudan, *Journal of Hydrology*, 409, 612–625.
16) Eltahir, E. A. B. (1996) El Niño and the natural variability in the flow of the Nile River. *Water Resources*, 32, 131–137.
17) Emanuel, K. (1991) A scheme for representing cumulus convection in large-scale models, *Journal of the Atmospheric Sciences*, 48, 2313–2335.
18) Flaounas, E., Bastin, S., and Janicot, S. (2011) Regional climate modelling of the 2006 West African monsoon: sensitivity to convection and planetary boundary layer parameterisation using WRF, *Climate Dynamics*, 36, 1083–1105.
19) Fontaine, B. and Janicot, S.:Wind-field coherence and its variations over West Africa, *Journal of Climate*, 5, 512–524, 1992.
20) Gallée, H., Moufouma-Okia, W., Bechtold, P., Brasseur, O., Dupays, I., Marbaix, P., Messager, C., Ramel, R., and Lebel, T. (2004) A high resolution simulation of a West African rainy season using a regional climate model, *Journal of Geophysical Research*, 109, D05108, doi:10.1029/2003JD004020.
21) Giorgi, F., Coppola, E., Solmon, F., Mariotti, L., Sylla, M. B., Bi, X., Elguindi, N., Diro, G. T., Nair, V., and Giuliani, G. (2012) RegCM4: model description and preliminary tests over multiple CORDEX domains, *Climate Research*, 52, 7–29, doi:10.3354/cr01018.
22) Gissila, T., Black, E., Grimes, D. I. F., and Slingo, J. M. (2004) Seasonal forecasting of the Ethiopian summer rains, *Interntaional Journal of Climatology*, 24, 1345–1358.
23) Grell, G. A. (1993) Prognostic evaluation of assumptions used by cumulus parameterizations, *Monthly Weather Review*, 121, 764–787.
24) Holton, J. R., Dmowska, R., and Philander, S. G. (1989) El Niño, La Niña, and the southern oscillation, Academic press.
25) Holtslag, A. A. M., De Bruijn, E. I. F., and Pan, H. L. (1990) A high resolution air mass transformation model for short-range weather forecasting, *Monthly Weather Review*, 118, 1561–1575.
26) Huffman, G. J., Bolvin, D. T., and Adler, R. F. (2011) GPCP Version 2.2 Combined Precipitation Data set, WDC-A, NCDC, Asheville, NC.
27) Jenkins, G. S., Gaye, A. T., and Sylla, B. (2005) Late 20th century attribution of drying trends in the Sahel from the Regional Climate Model (RegCM3), *Geophysical Research Letters*, 32, L22705, doi:10.1029/2005GL024225.
28) Kiehl, J. T., Hack, J. J., Bonan, G. B., Boville, B. A., Briegleb, B. P., Williamson, D. L., and Rasch, P. J. (1996) Description of the NCAR Community Climate Model (CCM3), NCAR Technical Note NCAR/TN-420+STR, doi:10.5065/D6FF3Q99.

29) Koteswaram, P. (1958) The Easterly Jet Stream in the tropics, *Tellus*, 10, 43–57.
30) Mitchell, T. D., Carter, T. R., Jones, P. D., Hulme, M., and New, M. (2004) A comprehensive set of high resolution grids of monthly climate for Europe and the globe: the observed record (1901–2000) and 16 scenarios (2001–2100), Tyndall Centre for Climate Change Research Working Paper, 55, 25 pp.
31) Müller, W. A. and Roeckner, E. (2008) ENSO teleconnections in projections of future climate in ECHAM5/MPI-OM, *Climate Dynamics*, 31, 533–549.
32) Nikulin, G., Jones, C., Samuelsson, P., Giorgi, F., Sylla, M. B., Asrar, G., Baüchner, M., Christensen, O. B., Déqué, M., and Fernandez, J. (2012) Precipitation climatology in an ensemble of CORDEX-Africa regional climate simulations, *Journal of Climate*, 25, 6057–6078, doi:10.1175/JCLI-D-11-00375.1.
33) Osman, Y. Z. and Shamseldin, A. Y. (2002) Qualitative rainfall prediction models for central and southern Sudan using El Niño–Southern Oscillation and Indian Ocean sea surface temperature indices, *Interntaional Journal of Climatology*, 22, 1861–1878.
34) Osman, Y. Z., Shamseldin, A. Y., and Abdo, G. M. (2001) El Niño- Southern Oscillation and rainfall variability in central and southern Sudan, *Water International*, 26, 177–184.
35) Paeth, H., Born, K., Podzun, R., and Jacob, D. (2005) Regional dynamical downscaling overWest Africa: model evaluation and comparison of wet and dry years, *Meteorologische Zeitschrift*, 14, 349–368.
36) Pal, J. S., Eltahir, E. A. B., and Small, E. E. (2000) Simulation of regionalscale water and energy budgets- Representation of subgrid cloud and precipitation processes within RegCM, *Journal of Geophysical Research*, 105, 567–594.
37) Pal, J. S., Giorgi, F., Bi, X., Elguindi, N., Solmon, F., Rauscher, S. A., Gao, X., Francisco, R., Zakey, A., and Winter, J. (2007) Regional climate modeling for the developing world: the ICTP RegCM3 and RegCNET, *Bulletin of the American Meteorological Society*, 88, 1395–1409.
38) Reynolds, R. W., Smith, T. M., Liu, C., Chelton, D. B., Casey, K. S., and Schlax, M. G. (2007) Daily high-resolution-blended analyses for sea surface temperature, *Journal of Climate*, 20, 5473–5496.
39) Ropelewski, C. F. and Halpert, M. S. (1987) Global and regional scale precipitation patterns associated with the El Niño/Southern Oscillation, *Monthly Weather Review*, 115, 1606–1626.
40) Segele, Z. T. and Lamb, P. J. (2005) Characterization and variability of Kiremt rainy season over Ethiopia, *Meteorology and Atmospheric Physics*, 89, 153–180.
41) Seleshi, Y. and Zanke, U. (2004) Recent changes in rainfall and rainy days in Ethiopia. *Interntaional Journal of Climatology*, 24, 973–983.
42) Shukla, J., Marx, L., Paolino, D., Straus, D., Anderson, J., Ploshay, J., and Kalnay, E. (2000) Dynamical seasonal prediction, *Bulletin of the American Meteorological Society*, 81, 2593–2606.
43) Steiner, A. L., Pal, J. S., Rauscher, S. A., Bell, J. L., Diffenbaugh, N. S., Boone, A., Sloan, L. C., and Giorgi, F. (2009) Land surface coupling in regional climate simulations of the West African monsoon, *Climate Dynamics*, 33, 869–892.
44) Sun, L., Semazzi, F. H. M., Giorgi, F., and Ogallo, L. A. (1999) Application of the NCAR

regional climate model to eastern Africa. Part 1: simulation of the short rains of 1988, *Journal of Geophysical Research*, 104, 6529–6548.

45) Sylla, M. B., Dell'Aquila, A., Ruti, P. M., and Giorgi, F. (2010a) Simulation of the intraseasonal and the interannual variability of rainfall over West Africa with RegCM3 during the monsoon period, *Interntaional Journal of Climatology*, 30, 1865–1883.
46) Sylla, M. B., Coppola, E., Mariotti, L., Giorgi, F., Ruti, P. M., Dell'Aquila, A., and Bi, X. (2010b) Multi year simulation of the African climate using a regional climate model (RegCM3) with the high resolution ERA-interim reanalysis, *Climate Dynamics*, 35, 231–247.
47) Sylla, M., Giorgi, F., Coppola, E. and Mariotti, L. (2013) Uncertainties in daily rainfall over Africa: assessment of gridded observation products and evaluation of a regional climate model simulation, *Interntaional Journal of Climatology*, 33, 1805–1817.
48) Thorncroft, C. D. and Blackburn, M. (1999) Maintenance of the African easterly jet, *Quarterly Journal of the Royal Meteorological Society*, 125, 763–786.
49) Vizy, E. K. and Cook, K. H. (2002) Development and application of a mesoscale climate model for the tropics: Influence of sea surface temperature anomalies on the West African monsoon, *Journal of Geophysical Research*, 107, doi:10.1029/2001JD000686.
50) Wang, G. and Eltahir, E. A. B.: Use of ENSO information in medium-and long-range forecasting of the Nile floods, *Journal of Climate*, 12, 1726–1737, 1999.
51) Wu, M. L. C., Reale, O., Schubert, S. D., Suarez, M. J., Koster, R. D., and Pegion, P. J. (2009) African Easterly Jet: Structure and Maintenance, *Journal of Climate*, 22, 4459–4480.
52) Zaroug, M. A. H., Eltahir, E. A. B., and Giorgi, F. (2014) Droughts and floods over the upper catchment of the Blue Nile and their connections to the timing of El Niño and La Niña events, *Hydrology and Earth Systems Sciences*, 18, 1239–1249, doi:10.5194/hess-18-1239-2014.
53) Zaroug, M. A., Sylla, M., Giorgi, F., Eltahir, E. A., and Aggarwal, P. K. (2013) A sensitivity study on the role of the swamps of southern Sudan in the summer climate of North Africa using a regional climate model, *Theoretical and Applied Climatology*, 113, 63–81.

Made in the USA
Monee, IL
02 November 2026

67385381R00116